INFRARED
THE NEW ASTRONOMY

INFRARED

THE NEW ASTRONOMY

DAVID A. ALLEN

A HALSTED PRESS BOOK

JOHN WILEY & SONS
New York   Toronto

Published in the U.S.A. and
Canada by Halsted Press, a
Division of John Wiley & Sons, Inc.,
New York.

Library of Congress Cataloging in Publication Data
Allen, David A.
    Infrared, the new astronomy.

    "A Halsted Press book."
    Bibliography: p.
    Includes index.
    1. Infra-red astronomy. I.   Title.
QB70.A6   522'.6   75-16584
ISBN 0-470-02334-1

Printed by Biddles Ltd., Guildford, Surrey, England

CONTENTS

Preface   7
1   The birth of the new astronomy   9
2   Techniques   21
3   The formative years   40
4   The solar system   51
5   Cool stars   76
6   Hot stars   97
7   Young objects   116
8   Irtrons and island universes   140
9   Hidden sources   158
10 Development of the science   183
11 Prospects   195
Appendix I   The Planck function   198
Appendix II   Update   202
References   205
Index   224

ACKNOWLEDGMENTS

My wife Carol was a constant help at all stages of the writing of this book. Much of the necessary reading was performed at the first-class library of the Royal Greenwich Observatory. Dave Calvert did an excellent job of making the prints. And I benefited greatly from many discussions with a great number of unsuspecting colleagues. My thanks also go to the following for permission to use the photographs:

To the Hale Observatories for plates 5, 6, 12, 13, 14, 15, and 17.

To George Herbig and the Lick Observatory for plates 7, 8 and 10.

To Kitt Peak National Observatory for plate 9.

To Richard Shorthill and the D. Reidel Publishing Co. for plate 2.

To Jim Westphal and the Astrophysical Journal (University of Chicago Press) for plate 3.

To Ed Nay for plate 4.

PREFACE

When you set out to climb a familiar mountain there is no necessity to plan your route in intricate detail. You can consult maps and guide books to save your feet from unduly straying, and there will be a choice of routes at various stages of the climb so that you may select whichever takes your fancy at the time. When at length you stand beside the summit cairn you will have no excuse if the route you chose was not the one best suited to the prevailing conditions and to your temperament. If, however, your aim is to scale a virgin peak your planning must be careful and precise. You must weigh the possible approaches and endeavour to select the one most likely to prove fruitful. If the way is tough you may have to abandon it. And when you clamber up the last few feet of the summit rocks to survey the view from your new vantage, you have no idea whether the route you did push through really is the easiest. Nor does it matter.

In writing this book I have had to scale a virgin mountain. There are no books on the subject on which to model mine, no guides to point the direction. I have picked my own way through the labyrinth and this is the diary of my journey. The route I took was relatively conventional, or at any rate logical, as befits a first ascent; those who follow may choose the less orthodox approaches. The choice is, in fact, quite large.

The application of infrared techniques to astronomical observation is a recent innovation. It is also a difficult job better suited to physicists and engineers than to astronomers. The number of practising infrared astronomers is small and many of them have only recently emerged from the chrysalides of physics laboratories. Personalities feature more prominently in this subject than in the conventional branches of astronomy. I could have written a book about these personalities. I could have told of bitterness and rivalry and of excessive competition. I could have shown how different groups, starting as good friends, became so embittered that they would not even acknowledge one another's publications when writing papers. I could have made much of the abysmal quality of many of the earlier papers: of hasty and inconsistent observations, of overinterpretation of data, of the many errors that appeared in print. I could have narrated clandestine observing malpractices and even the resultant loss of financial support from a government agency. Because the subject is young, such occurrences, while not pardonable, are hardly unexpected. But it would be churlish to make them the substance of a book about infrared astronomy, at least until time has healed the deeper scars.

Because infrared touches all aspects of this copious science, from our parochial satellite to the farthest quasar, this book covers a wide range of topics. It was my intention that it should interest the scientist with a fairly broad knowledge of basic astronomy and the specialist more deeply concerned with one facet of astronomy. The inclusion of an extensive list of references should provide anyone with all the information he needs to delve deeper into the subject.

In the last decade infrared astronomy has established itself as one of the most important tools of the science. As a result of the observations made at infrared wavelengths and described in this book we have had to reformulate our thinking on a great variety of astronomical topics. Important papers detailing unanticipated discoveries have festooned the pages of astronomical journals: this has been an exciting decade.

No complete review of the advances in infrared astronomy exists. While a comprehensive treatment has long been overdue, potential authors have been dissuaded from writing one by the rapid growth and capricious nature of the subject. In the last few months, however, there have been unmistakable signs that the surge of discovery is slowing and that the time is now right for the publication of a book which will not too quickly become out of date.

The main text of this book was finalised on 20 June 1974. Papers which I had not seen by that date were not included. In an attempt to render the book as up to date as possible, appendix II has been added at the final proof stage (early 1975) and contains a digest of the more significant developments of the intervening months. As I write this preface, eight weeks after my deadline, only two papers have appeared in print which contain new data worthy of inclusion in appendix II.

Hailsham, Sussex, England.
August 1974

8

1
THE BIRTH OF THE NEW
ASTRONOMY

Of the great wealth of astronomical texts published in recent years, few indeed deign to mention the feasibility of astronomical observation at infrared wavelengths. Optical and radio results fill tomes, but the tract of the electromagnetic spectrum bridging their domains is treated as but a poor relation. The reason is not hard to find: along with other branches of astronomy exploiting the more esoteric reaches of the spectrum – X-ray and ultraviolet for example – the principal advances and discoveries have all been made within the last decade. Infrared is one of the new astronomies, and is only now beginning to claim its share of the limelight. And yet the roots of infrared astronomy reach back to a time long before Jansky discovered radio signals from the Galactic centre, before most of the world's major optical observatories were even conceived, to the time when astronomy was in the hands of the amateurs.

Infrared – a term used loosely to describe the infrared portion of the spectrum – was discovered at the end of the eighteenth century by Sir William Herschel. Herschel was concerned with the amount of heat transmitted through the coloured filters he employed when observing the sun. Some filters, he found, cut down the light considerably but allowed into his eyes too much heat for comfort; other filters had the reverse effect. He thought this to be a function of the colour of the filter and therefore suspected that light of different colours had different heating powers. The experiment he devised to investigate this was both simple and effective. He passed sunlight through first a slit and then a prism to produce a solar spectrum in the then conventional way. He then arranged three thermometers so that the bulb of one could be placed in the different colours of the solar spectrum while the others acted as controls. After a few minutes, when equilibrium had been reached, he noted the readings of each thermometer and found that the highest temperature was recorded when his thermometer lay in red light. Acting apparently on a hunch, he placed one thermometer beyond the red

Fig. 1 Herschel's experiment to detect infrared radiation. One thermometer
is placed in the infrared radiation; the other two act as controls.

end of the visible spectrum and found it to read higher than the two controls (see Fig 1). Herschel had detected energy in the infrared part of the spectrum; he had in effect observed the sun at infrared wavelengths.

In a series of four papers to *Philosophical Transactions* in 1800[1] Herschel described his discovery and subsequent investigation of what he then called calorific rays and later became known as infrared radiation (an odd term suggesting that such radiation lies somehow *beneath* visible light). He found that the calorific rays were reflected, refracted, transmitted and absorbed just like visible light, and it became clear that he was dealing not with another property of light, but with an extension of it into a previously unexplored spectral region.

Herschel had done no more than scratch at the surface, but his work set the scene for what was to become one of the great break-throughs of the century – the opening up of the electromagnetic spectrum. Fig 2 illustrates the spectrum as we now know it. In Herschel's day only the tiny section marked *visible* had been explored, the section to which our eyes are sensitive and in which most energy is available to earth-based beings. Unlike the visible, the infrared is not a well-defined portion of the spectrum; the term is commonly used to describe the huge amorphous chunk between the far red at about 7,000 Å and the radio at a few millimetres wavelength. Throughout this range, wavelengths are usually measured in *microns*, written μ. More properly this should be *micrometres*, μm, but although the latter symbol is often used in print, the longer term is never used in speech.

Visible light has wavelengths between 0·4 and 0·7 μm. From 0·7 to about 1·1 μm lies a region I shall call the photographic infrared. Throughout this waveband, photographic emulsions and the photo-cathodes employed in image tubes are sensitive. All the techniques of optical astronomy other than simple visual observation can therefore be employed. Beyond 1·1 μm where photographs cannot be taken, we enter a new domain – the strange world of the infrared. The long wavelength boundary is usually set at 1 mm, or 1,000 μm. At about this wavelength the detection techniques of radio astronomy are more suitable than those employed by infrared astronomers.

Three orders of magnitude in wavelength is too great a range for one word to embrace. To specify smaller sections of the range sub-divisions were introduced by the addition of prefixes near-, mid- and far-. It would seem natural to allocate these prefixes to the three decades covered, so that for example the mid-infrared referred

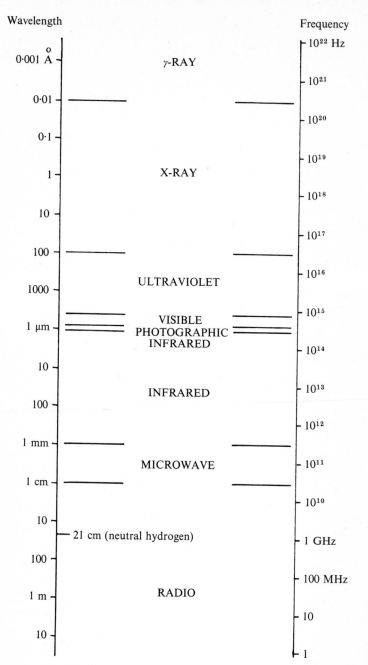

Fig. 2 The electromagnetic spectrum.

to the wavelengths 10 to 100 μm. No such logic prevailed, however, and the terms have been so variously applied as to make a mockery of the system. The optical spectroscopists were the most frequently guilty in this respect: to them 0·8 μm was 'far-infrared'. In the present book a rather different system will be employed defined by the techniques of given wavelengths rather than by the wavelengths themselves. Hence the introduction of the term photographic infrared above.

Infrared astronomers think mostly in terms of wavelengths, but an exactly equivalent system could be defined in terms of the frequency of the radiation. Those weaned on radio astronomy often find it more convenient to think in frequencies; Fig 2 has therefore been equipped with a frequency scale.

Although our eyes do not see in the infrared, our bodies are sensitive to radiation at surprisingly long wavelengths. The human nervous system responds to visible and infrared radiation alike by absorbing it and by warming up. Infrared can therefore be thought of as radiant heat; it is the energy radiated by cool objects. As soon as the element of an electric fire is switched on, a hand held near it can detect the infrared it radiates. As the element gets hotter, so the radiation emitted increases. Eventually it begins to glow a dull red, and then brighter orange. It is now emitting some of its radiation in the visible. If we could continue to increase the temperature of the element without burning it out, it would glow yellow and eventually white hot. The hotter an object, the more its radiation is weighted towards the blue end of the spectrum; the cooler the redder. This is the principle exploited by the optical pyrometer, a device which determines the temperature of an inaccessible object merely by its colour. A pyrometer cannot measure a temperature lower than about 1,200 °K (900 °C) because the radiation is then emitted solely in the infrared.

BLACK BODIES

If we measure the energy radiated by an object at a number of different wavelengths we get an *energy distribution* which we can plot graphically. Ideally this will be the Planck function, or black body curve. The Planck function is a theoretical energy distribution deduced from quantum physics. It is described in mathematical form in Appendix 1 which, essentially, parallels this section of Chapter 1.

Basically the Planck function relates the radiant energy flux, F, to the wavelength of observation, λ, and the temperature, T, in

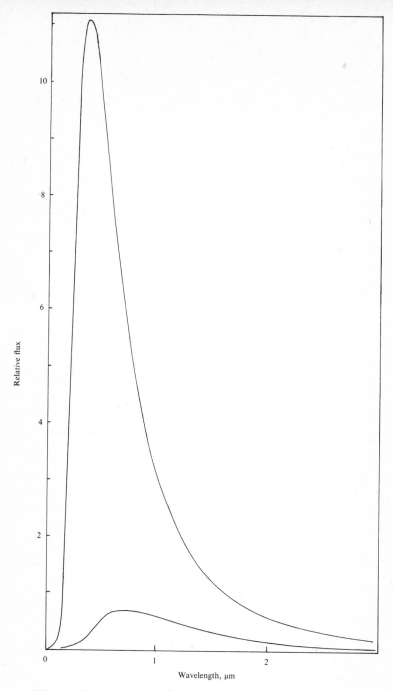

Fig. 3 The Planck function for black bodies at 10,000 and 5000 °K.

degrees Kelvin of the object concerned. F is also proportional to the apparent area of the object. When plotted on a graph of F against λ the Planck function is a family of curves each defined by the parameter T. Two of these curves, with T = 10,000 °K and 5,000 °K, are plotted in Fig 3. If we use logarithmic axes the Planck function has the form shown in Fig 4. Logarithmic coordinates are very useful because the Planck function, when plotted on them, retains the same shape whatever the value of T. Throughout this book energy distributions of astronomical objects will be plotted on logarithmic graphs with fluxes measured in Watts per square centimetre and wavelengths in microns. The actual energy we receive from the source depends on the area of our telescope mirror (in cm²), and it should be noted that these fluxes are the values that would be measured from above the

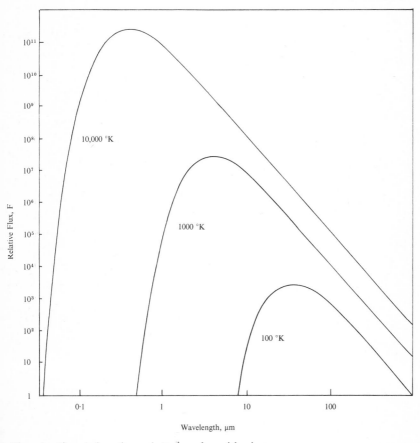

Fig. 4 Planck functions plotted on logarithmic axes.

earth's atmosphere; they have been corrected for the effects of transmission to ground level.

While the shape of the Planck function does not change with temperature, its position does. This motion is expressed in two laws, Stefan's and Wien's. Both laws are derived in the Appendix and will merely be stated here.

Stefan's Law: vertical displacement on Figs 3 and 4

Stefan's law states that the total energy which is emitted by an object over all wavelengths is proportional to the fourth power of its absolute temperature. Thus:

$$E = \sigma a\, T^4$$

where we have written σ for the constant of proportionality, Stefan's constant. The other factor in this equation, a, is the surface area of the object. Obviously, the larger a body the more energy it can radiate.

Fig 3 illustrates Stefan's law: the 10,000 °K body curve rises to a point sixteen times higher than the 5,000 °K curve. This rapid change in the emitted flux is not always so apparent on logarithmic plots. Neighbouring curves on Fig 4 differ by a factor of ten thousand in their emitted fluxes.

Wien's Law: horizontal displacement on Figs 3 and 4

As the absolute temperature falls, the black body curve on Fig 4 moves to the right. This displacement in wavelength is inversely proportional to the temperature, T. It is simplest to choose a specific point on the black-body curve to describe its motion along the wavelength axis: the most obvious choice is the peak of the curve, and we shall call the corresponding wavelength λ_{max}. Wien's displacement law is simply stated as:

$$\lambda_{max}\, T = \text{constant.}$$

The value of the constant is approximately 3,700 μm K; thus a 1,000 °K black-body curve peaks at a wavelength of nearly 4μm. Wien's law is of unparalleled importance in infrared astronomy for it enables us to deduce the temperature of an object by observing its energy distribution. Once we know the temperature we can use Stefan's law to determine the only remaining free parameter, the size of the object; but in practice a rather large number of assumptions are involved in this last step, and it is rarely possible to deduce a reliable size.

The Rayleigh – Jeans Tail

For many purposes we can divide the black-body curve into two portions according to whether λ is longer or shorter than λ_{max}. When λ exceeds λ_{max} we are on the Rayleigh–Jeans tail of the curve. Here the flux merely falls as the inverse cube of the wavelength. On Fig 4 this is a straight line of gradient -3. On the Rayleigh–Jeans tail the shape of the energy distribution is no longer dependent on the temperature of the object, and we cannot use Wien's law to deduce that temperature. We shall most often be concerned with the Rayleigh–Jeans tail when observing stars in the infrared. For stellar temperatures of 5,000 °K or more, λ_{max} lies shortwards of 1 μm, and all measurements in the infrared lie on the Rayleigh–Jeans tail.

The leading edge of the Planck curve, where $\lambda < \lambda_{max}$, rises extremely steeply. Were we to observe a 1,000 °K object at 1 and then at 1·5 μm, we should find the signal at the shorter wavelength twenty-four times weaker, and quite possibly below our detection level. It is the existence of this very steep rise in the Planck function which gives infrared astronomy its special appeal, for there are cool sources in the sky which are extremely prominent at long wavelengths and yet are totally unsuspected at shorter. If we never make observations at wavelengths longer than, say, 5 μm we will never know of the existence of objects cooler than about 200 °K. By extending the bounds of our knowledge further into the infrared we progressively open up a totally different world until finally we reach a limit somewhere about 1 mm – the farthest bound of the infrared region. At this limiting wavelength the bodies we might hope to observe are so cold that they do not radiate enough energy for our insensitive equipment to detect. If there are celestial objects at temperatures of a few degrees Kelvin, we may never know of them.

Emissivity

The Planck function is a mathematical concept which describes the ideal behaviour of black bodies. A black body is defined, not very helpfully, as one which obeys Planck's law. While black bodies are found in fairy tales and scientific papers, the real world contains no such thing. Some celestial bodies approach quite closely the quality we may call blackness, and in the laboratory the nearest approximation is afforded by a small aperture in a large box whose inner walls are covered with matt black paint.

The degree of divergence from blackness is *emissivity*, ε, which is simply a scaling factor acting on the Planck curve. ε is unity for a true black body and cannot exceed this value. An ε of less than one defines another mythical beast occasionally found in the literature: the grey body. In practice all objects have an ε which varies with wavelength, and sometimes these variations are so great that the resulting energy distribution of the object may bear little resemblance to that of a black body. In many astronomical applications ε falls with λ throughout the infrared region, but examples in which the emissivity is very small, except within a narrow range of wavelengths, will also be met in the later chapters of this book.

THE INFRARED WORLD

Suppose our eyes were sensitive not to visible light but in the infrared, at a wavelength of 10 μm: how then would the world look? It would be dominated not by sunlight, as now, but by the thermal radiation of room-temperature objects. A rock in sunlight would not necessarily be brighter than one in shade; only after absorbing sunlight for a few minutes would it warm up enough to become brighter. The entire sky would be bright, day and night, rendering stars invisible. Animals, being warm, would appear bright; inanimate objects darker. A centrally heated house would be brighter than an unheated one. Inside a refrigerator it would be too dark to find anything.

Objects dark at optical wavelengths are dark because they absorb most of the incident light, but in so doing they become warmer than white objects, and hence in the infrared brighter. In our infrared world strange reversals would result whereby a 'white' shirt could be darker than the 'black' suit worn with it. Brighter spots would be produced on the suit by holding a hand against it for a few seconds. We could even judge the temperature of our bath water by its brilliance.

It is the fact that living animals stand out brightly against their background, even at night, that gives the 10 μm region its practical importance, especially in military applications. Nature too has not neglected the fact, wisely equipping certain snakes with very sensitive infrared detectors for the location of their prey at night.

In the nineteenth century, however, the military had neither the interest nor the ability to produce any detectors more sophisticated than Herschel's thermometer. Astronomers, being unwilling to train snakes to look through telescopes, had to await the work of physicists in improving detection systems. Improvements were slow to come. From the thermometer to the thermocouple to the vacuum thermopile seem but small steps to us but were major developments then. The thermopile is no more than a number of thermocouple junctions united in one cell, and it was with these that the pioneer infrared astronomers detected radiation from celestial bodies.

The sun was easy – even at wavelengths beyond 1 μm. The moon, the next brightest astronomical source, was less so. An Italian observing from Vesuvius claimed the first detection, but credit for the first conclusive measurement is usually given to the Scottish Astronomer Royal, Piazzi Smyth.[2] The relevant observations were made during his 1856 expedition to the island of Tenerife (it is no mere coincidence that this same site has been chosen for Britain's first infrared telescope). Piazzi Smyth used a thermocouple and was just able to detect the full moon. By 1870, however, the 4th Earl of Rosse (the son of the famous 3rd Earl who built the 72-inch telescope) had been able to measure the moon through its phases.[3] He deduced the range of its surface temperature to be 500 °F. The currently accepted value is 300 °K (540 °F), and Rosse's results are all the more impressive when it is realised that Stefan's law had not at the time been formulated.

The technique employed by astronomers in the first half of this century was to measure a source both with and without a filter in the beam. At first a block of glass was used as the filter material; later the water cell was introduced, a layer of water trapped between two thin glass slides. By application of Wien's law, the ratio of the signals yielded a measure of the temperature of the source – although the value was subject to large uncertainties. In the early part of the twentieth century the moon, the principal planets and the brighter stars were thus measured by pioneers in the USA.[4] Few unexpected results emerged from this work, and many of the astronomical papers published then contain mostly descriptions of the apparatus and only a minimum of quantitative data.

One interesting result that did emerge concerned the moon. Pettit and Nicholson were the lunar experts. Using the Mount Wilson 100-inch telescope they measured the temperature of small areas of the surface during the lunar day and in eclipse.[5] The results were surprising: in eclipse the moon cooled more than expected. The theoreticians promptly set to work. Their job was to solve a differential equation describing the heat flow at the lunar surface. The equation, because of the σT^4 term of Stefan's law, did not admit an analytical solution. Instead, laborious numerical calculations had to be made, calculations that take seconds on a modern computer but required weeks of work then. The man with the necessary patience was Wesselink, and in 1948 he published the results of his computations.[6] The moon, said Wesselink, is covered not with rock

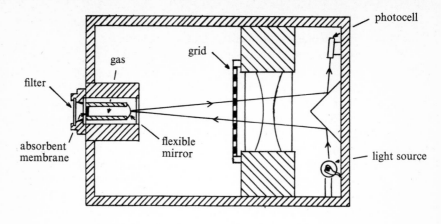

Fig. 5 The Golay cell. Energy absorbed by the membrane causes the gas to expand which changes the curvature of the flexible mirror. This in turn alters the amount of light which makes the double pass through the grid between the source and the photocell.

but with a fine powdery dust. That he was right we now know full well, but at the time this unanticipated finding was both controversial and an ample demonstration that infrared astronomy was more than the mere idle doodlings of eccentric astronomers who should have been employed in taking spectra of every star in the sky.

But these first results soon became entangled in the folded web of time. The pioneer papers lay gathering dust in the great libraries of the world. All the useful measurements that could be taken with the primitive thermopile had been taken. Infrared astronomy had, it seemed, had its day. Until the present revolution, that is.

In 1947 Golay published details of a subtle pneumatic detector for infrared radiation.[7] Here was the first release from the strait-jacket world of thermopiles and water cells. Golay's cell is delightfully anachronistic – the sort of thing James Watt might have invented – yet amazingly sensitive. The principle is shown in Fig 5. Radiation entering the cell through a transparent window heats the gas within. As the gas expands it causes a rubber diaphragm to bow outwards, and a light beam reflecting off this diaphragm is deflected. When equilibrium is reached the deflection of the light beam is a function of the radiation entering. Golay cells are used extensively in laboratory infrared work, but astronomers were lax in making use of this new tool, and by the time their interest was aroused, better detectors had been devised, detectors which were to revolutionise the field and lead at last to the discoveries this book will describe.

2
TECHNIQUES

Photographic emulsions can be made to respond at wavelengths as long as 1 μm. To this radiation they are, however, very insensitive. Photocathodes – surfaces which release electrons when radiation strikes them – can now be made to respond beyond 1 μm, and even to 1·5 μm; fibre-optic image tubes employing these photocathodes can be used to take astronomically useful photographs at about 1·1 μm. Thus we have the potential of knowing what the sky looks like at 1 μm, although only a small fraction of it has yet been sampled. This is the photographic infrared, the Herschel region. Few startling or unanticipated results have emerged from work in the photographic infrared and only occasional mention of it will be made throughout this book. This chapter concerns the techniques of present-day infrared astronomy and refers solely to the region longward of 1·1 μm.

Since no suitable photographic or other image-producing device is yet available, all infrared measurements must be made by *photometry*. As at optical wavelengths, photometry consists of focusing onto a single detector a small patch of sky (containing the star or other object of interest) and of measuring the energy radiated by that patch of sky in the particular waveband chosen. There is a wide range of detectors from which to select. For the last few years two have proved to be of superior sensitivity; these are the lead sulphide (PbS) cell and the doped germanium bolometer.

LEAD SULPHIDE

Many people are familiar with the photographic exposure meter containing a cadmium-sulphide cell. Light falling on this photosensitive material causes a physical change in the crystal which manifests itself by a change in the resistance of the cell. A small battery passes a current through the cell and through an ammeter; as the resistance of the CdS changes, so the current is altered and the ammeter reads its value, a value which is a measure of the incident radiation.

The PbS cell works in exactly the same way, the only difference being that it is sensitive to radiation of from 1 to 4 μm wavelength. Thus, 1 to 4 μm is the portion of the infrared I shall call the PbS region. Manufacturers of lead-sulphide cells can produce them to various specifications. The principal variants are the desired sensitivity, the operating temperature and the time constant for the cell to respond to changes in the incident radiation. Infrared astronomers always seek the greatest possible sensitivity, and are not too fussy if it takes the cell a tenth of a second or so to realise that radiation is hitting it. For extremely high sensitivity, however, one has no choice but to cool the cell considerably below room temperature. Even solid CO_2, used in most optical photometers, is barely chill enough, and liquid nitrogen is normally employed as the coolant. Liquid nitrogen is not difficult to work with: it is cheaper than beer and almost as easily available; it pours like water and can be kept for hours in a normal thermos flask; yet it boils at a temperature of 77 °K (-196 °C).

DOPED GERMANIUM CELLS

Beyond about 4 μm the energy carried by individual photons is too low to effect a change in PbS; other detectors are therefore needed. In the 1950s it was realised that certain semiconductors have the desired properties. The base material for most semiconductors is the strange half-metal germanium. In normal metals the electrons move freely and allow the conduction of electricity; in the germanium crystal the electrons are loosely bonded and can be freed only by certain external stimuli. By introducing traces of impurities, a process known as *doping*, the crystals can be 'tuned' to respond to the desired stimulus – in this case to infrared radiation. If copper or mercury are the impurities, the germanium crystal becomes conductive under the influence of radiation in the 5 to 15 μm region. Such crystals must, however, be cooled far below liquid nitrogen temperatures to about 10 or 20 °K. Only two cryogenic liquids are readily available whose boiling points lie in this range – liquid neon and liquid hydrogen. The former is prohibitively expensive; the latter is avoided by the more timid because of its tendency to explode. Ge:Cu and Ge:Hg cells are usually cooled in liquid helium, to a lower temperature than is strictly necessary. Liquid helium boils at 4·2 °K, and at this temperature all other substances are frozen solid: it is the coldest known liquid, is therefore expensive to make and difficult to handle. A thermos flask of it would boil away in seconds. It cannot be poured from vessel to vessel but must be transferred under pressure through specially constructed insulated tubes, and then only with considerable losses.

For most applications the Ge:Cu and Ge:Hg cells have been superseded by Ge:Ga. When germanium is doped with traces of gallium its electrons can be freed by radiation of any infrared wavelength. A detector equally sensitive to all wavelengths is known as a *bolometer*. The Ge:Ga bolometer was invented in 1961 by Frank Low, then a physicist at the Texas Instrument Corporation, and it is often referred to as the Low bolometer.[1] Not only does the Low bolometer function over a much greater wavelength range than Ge:Cu or Ge:Hg, but additionally for conventional infrared photometry its sensitivity is superior by up to an order of magnitude in even the 5 to 15 μm region.

All this gain inevitably means some sacrifice elsewhere, and indeed there is a great sacrifice in convenience. Not even liquid helium boiling at 4·2 °K is cold enough for Low's bolometer: it must work at 2 °K or less. To achieve such low temperatures liquid helium is contained in a continuously evacuated enclosure, for by lowering the pressure, the temperature at which it boils is reduced. A powerful vacuum pump is therefore a standard piece of the infrared astronomer's equipment.

Although there are some minor differences between the techniques in the PbS and doped Ge regions, the basic principles of infrared photometry apply equally to both, and these will now be discussed with reference to the differences where necessary.

THE ATMOSPHERE

Without its atmosphere planet earth would be a barren lifeless place indeed. There are many astronomers, however, who would give their eye teeth to be able to switch off the restless gases that perpetually seethe and shimmer before their telescopes. Infrared astronomers share this desire, for the earth's atmosphere offers as great a handicap to them as Machiavelli could have devised. Intolerant and unyielding, the air we breathe absorbs all but a tiny portion of the infrared photons that have journeyed laboriously from the far reaches of the universe. So long as we keep our feet planted on *terra firma*, we have no choice but to conform to Nature's whims and glean what little we can from the meagre sample. The principal actors in this melodrama are the two molecules CO_2 and H_2O, mostly the latter. Throughout the infrared they have broad and deep absorption bands of sufficient strength to absorb totally all radiation of the appropriate wavelengths. Only at wavelengths in the narrow gaps between the molecular bands – the so-called atmospheric windows – can radiation penetrate to ground level relatively unimpeded. Fig 6 illustrates the infrared windows, and their properties are listed in Table 1. The windows beyond 10 μm are very opaque and were not known to the early infrared astronomers. The first 20 μm observation (of the sun) was made in 1942.[2]

In the visible and photographic infrared, narrow portions of the spectrum have been selected to define standard photometric systems. The most commonly used is that of Johnson, and his wavebands (and their defining filters) are designated by the letters UBVRI (ultraviolet, blue, visible, red, 'infrared'). Most of the infrared atmospheric windows are sufficiently narrow essentially to define photometric bands without the use of filters. Interference filters are usually used to sharpen their profiles, however. When Johnson's system was extended into the infrared the lettering scheme was continued. Thus the 1·25 μm window became known as J, the 2·2 μm window K and so on. Only subsequently was it realised that an equally useful window at 1·65 μm had been omitted, and this was eventually labelled H for no good reason. The 10 μm window is so wide that it is frequently subdivided. The letter N referred to the entire window, but narrower sections were designated O and P. These early subdivisions have been abandoned; the present trend is for individual observers or groups of observers to choose their own restricted wavebands, and no internationally accepted photometric system is evolving. Identification by lettering has become impractical; instead the wavebands are coded by the much more sensible system of putting the mean wavelength in square brackets, thus: [8·6 μm],

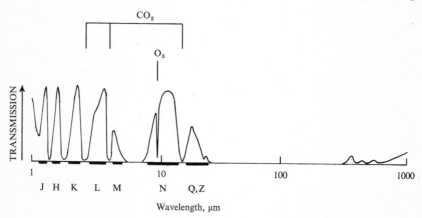

Fig. 6 The atmospheric windows accessible from the earth's surface.

[12·2 μm]. On the other hand, the PbS photometric bands have been universally adopted, and J, H, K and L are in constant use in the literature. They will be used here. The 5 μm band, although widely adopted, is more often referred to as [4·8 μm] than M. Some observers use a narrow portion of the 3·5 μm window, from 3·6 to 4·0 μm, and refer to it as [3·8 μm].

TABLE 1

TABLE 1

The Infrared Atmospheric Windows

waveband, μm	mean wavelength	best detector	photometric letter	typical sky transparency	sky brightness
1·1–1·4	1·25 μm	PbS	J	high	low at night
1·5–1·8	1·65	PbS	H	high	very low
2·0–2·4	2·2	PbS	K	high	very low
3·0–4·0	3·5	PbS/InSb	L	3·0–3·5 μm fair 3·5–4·0 μm high	low
4·6–5·0	4·8	InSb/Ge:Ga	M	rather low	high
7·5–14·5	10	Ge:Ga	N	8–9 and 10–12 μm fair, rest low	extremely high
17–40	20 35	Ge:Ga	17–25 μm (Q) 28–40 μm (Z)	very low	very high
330–370	350	Ge:Ga/Si		extremely low	low

Usually the windows are sufficiently narrow that one may refer to a single discrete wavelength, the *effective wavelength*. Whenever a wavelength is used in this book it will mean the effective wavelength of the window or waveband. For wide bands, however, the effective wavelength is a function of the temperature of the object observed. A cool object weights more heavily the longer end of the atmospheric window, and thus increases the effective wavelength of the observation.

The absorption of radiation is not the only difficulty the atmosphere introduces. In addition the sky emits strongly in the infrared. The energy distribution of the atmospheric thermal emission is that of a body at about 270 °K modified by its emissivity. The emissivity is essentially unity in the absorption bands, but may be as little as a few per cent in the clearest atmospheric windows. Even when the emissivity is low, the energy received from the sky at the detector can be very large compared to the energy of the source being examined. Highly efficient discrimination is therefore needed to distinguish the radiation of the source from that of the sky. The problem is most acute in the 10 μm band where a 270 °K black body peaks. At J, H and K, atmospheric thermal radiation is negligible and only scattering of radiation by small particles can cause the sky to be bright. Scattering becomes rapidly less efficient the longer the wavelength, which is why the sky is blue not yellow; at 2·2 μm the sky is quite dark by day and by night.

Nor is the atmosphere the only thing radiating at 10 μm. All room-temperature objects do so, and this includes the mirrors of the telescope, the supports to the secondary and any other items that are seen by the detector. Observing at 10 μm has been likened to looking for stars in daylight with a luminous telescope. All this stray radiation would not be an insuperable problem if it were constant. Alas, the temperature of the sky – and to a lesser degree of the telescope – undergoes fluctuations with periods from hours to microseconds. Minute though these fluctuations are, they are easily detected by a Low bolometer and give rise to a varying signal which has been called *sky noise*. All detectors have inherent electrical noise which dictates their ultimate sensitivity. Ideally one wants to reduce any extraneous sources of noise (eg that caused by vibration) to a level below that of the detector. At 10 μm the sky noise exceeds the inherent noise of a Low bolometer at most points on the earth's surface. In choosing a site from which to make 10 μm observations, therefore, one must know how great the sky noise is. Unfortunately the factors which influence sky noise are not understood. It certainly correlates with the presence of cirrus clouds, and it is probably low in places with a very dry climate. More than that cannot at present be asserted with confidence, and selection of observing sites has largely been on a trial-and-error basis, so far with remarkably little of the error.

Birds and insects passing in front of the telescope also radiate and introduce spurious signals. More than one clear night has had to be abandoned because swarms of moths were flying round the telescope. This is clearly another factor to be considered in selecting an infrared site!

To reduce the signal, and hence the noise, from the telescope and sky it is desirable to use a very small beam, only a few seconds of arc across. Then, however, one is confronted with the problem of *seeing*. In optical photometry small beams are avoided when possible because the turbulence of the atmosphere can cause part of the energy of the star to fall outside the field of the detector. Infrared seeing appears to correlate only in part with optical seeing, but there are occasions when a source exceeds in size the small beams that must be employed. When this happens the signal fluctuates with the turbulence, with resulting loss of accuracy.

THE TELESCOPE

In general, any optical telescope which relies entirely on reflecting surfaces can be used for infrared photometry. A telescope which wholly or in part uses transmitting optics is unsuitable because glass absorbs infrared radiation. Most of the large reflecting telescopes in the world have been used for infrared work. All function well in the PbS region and rather less well at 10 μm. A few telescopes have been designed specifically for infrared use, and these will be described in the next chapter. Because water vapour is the principal cause of the atmosphere's absorption and thermal radiation, telescopes at high-altitude sites in very dry climates are usually the best for infrared work. The Cassegrain focus is invariably selected because of its convenience.

CHOPPING AND NODDING

All infrared photometry utilises a procedure known as chopping. Two neighbouring patches of sky (*beams*) are presented alternately to the detector and their difference is taken electronically by phase-sensitive rectification. The frequency of chopping between the two beams is chosen to suit the response time of the detector and may be anywhere between about 1 and 1000 Hz.

There are three reasons why chopping is used:

(i) It is much easier to amplify an alternating than a direct current.

(ii) Most detectors tire of seeing the same thing all the time; if the incident flux does not change the response of the detector may vary.

(iii) Chopping provides the necessary discrimination against a bright sky: the background, if identical in the two beams, cancels out and only the signal from a source located in one or other beam is recorded.

If there is a gradient of the sky brightness, one beam will have a larger background than the other and a spurious signal will result. To eliminate this possibility the following procedure is adopted:

1. Set the telescope so that the source is in one beam (the beam treated as positive during rectification) and measure the signal by integration. This will comprise two parts – the signal from the source, S, and the difference between the two beams, Δ.

2. Move the telescope so that the source is in the other beam (where the incident flux is treated as negative) to produce a signal $-S$ from the source but still Δ from the background.

3. Subtract the two signals to get $S + \Delta - (-S + \Delta) = 2S$.

The process of driving the telescope back and forth between the two beams is known as *nodding*. For typical integration times the telescope must move every 20 to 30 seconds. At some observatories

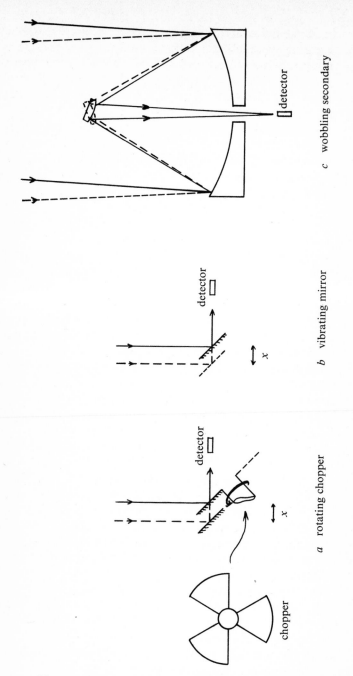

Fig. 7 Three popular modes of chopping infrared radiation.

this must be done manually by means of slow-motion drives; at more sophisticated establishments the telescope is nodded automatically. The separation of the two beams is usually in the range 6″ to 1′ arc. It must clearly be larger than the beam diameter, but should also be as small as possible to ensure that the background signals in the two beams are closely related.

There are three mechanical techniques one might consider to produce the chopping. One would be to move the telescope back and forth as for nodding. This is clearly impractical at the required frequencies. Another would be to vibrate the detector back and forth across the focal plane of the telescope. This, however, introduces microphonics (vibration noise). Therefore it is necessary to introduce a moving mirror into the light path which has the effect of moving the focal plane of the telescope back and forth in front of the detector. Fig 7 illustrates schematically the usual techniques. Method (a) involves a rotating segmented mirror. When a segment is in front of the detector, radiation reaches it down the right-hand path; when a gap between segments passes in front of the detector the radiation is reflected off a fixed mirror via the left-hand path. Two different parts of the focal plane separated by the linear distance x are alternately sampled. The rotating chopper is the most efficient method and is usually employed in the PbS region. At longer wavelengths, however, it has two disadvantages: first the two mirrors may vary in temperature to produce spurious signals; and second the edges of the segments, which are inevitably subject to minor flaws and irregularities, emit strongly and may reflect warm objects. As a practical point it is extremely difficult to manufacture and mount an optically flat segmented mirror; good specimens are much-prized possessions and to break one is a cardinal sin.

Method (b) achieves exactly the same effect by moving a mirror back and forth at the desired frequency. The mirror may be driven mechanically – as by a rotating cam – or electrically. Its efficiency is lower than that of the rotating chopper because time is wasted as the mirror moves between its two positions. If the chopping frequency is higher than about 20 Hz this becomes a serious disadvantage. At long wavelengths the vibrating mirror is not subject to the disadvantages of the rotating chopper. However, at 10 μm some thermal noise is still introduced because the two beams to the detector include slightly different portions of the telescope optics. This can be further reduced by method (c), in which the chopping component is the Cassegrain secondary tilted back and forth. A wobbling secondary allows one reflection less, and since 20 per cent of the incident signal may be lost at each reflection this is a major saving. Many existing telescopes do not readily lend to the construction of a wobbling secondary, but the option is now considered at the design stage of any telescope on which infrared observations may be made. For very large telescopes it is generally impractical because of the sheer size of the secondary mirror.

Because in infrared astronomy we can examine the contents of only one beam at a time, mapping of an extended object or searching for infrared sources is a difficult business. It is usually accomplished by raster scanning the telescope back and forth. Difficulties arise because one is continually sampling the difference between two adjacent beams. When searching for a source one may raster scan along the line of chopping or at an angle to it. In the former a source will appear as a double deflection of the signal, one of each sign as it passes in turn through the two beams. In the latter the source may be detected in either beam. Mapping an object bigger than the beam separation is an exacting task becaues of the differencing procedure involved.

THE PHOTOMETER

The portion of an infrared astronomer's equipment which is attached to the telescope comprises two basic items: a photometer and a dewar. Designs of both parts vary considerably. The diagrams refer to a PbS detector and the design should be regarded merely as illustrative.

The photometer is the piece that contains the chopping mechanism and the eyepieces. It might be thought that only a simple lightweight box with a couple of mirrors would suffice, but this is not the case. The photometer must be rigidly constructed and contain an assortment of gadgets; a typical example would weigh 50 lb.

Fig 8 shows a very simple photometer for use with a straight-through Cassegrain telescope. The photometer is bolted onto the mounting plate of the telescope immediately behind the primary mirror, and light enters directly from the secondary. It is necessary to guide the telescope whilst making the measurement, and this is best accomplished by using a beamsplitter, a device which transmits visible radiation to an eyepiece but reflects the infrared to the detector. The best beamsplitters are made from thin gold films coated with zinc chloride. Not all infrared sources are visible objects, however, and sometimes it is necessary to offset guide on a nearby bright star. To do so one must move the guiding eyepiece relative to the

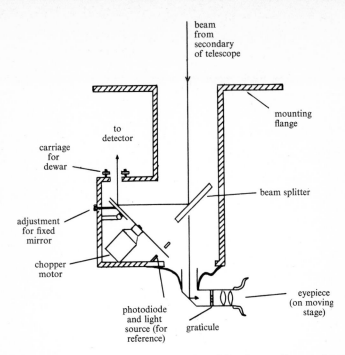

beam
from
secondary
of telescope

mounting
flange

to
detector

carriage
for
dewar

beam splitter

adjustment
for fixed
mirror

chopper
motor

eyepiece
(on moving
stage)

photodiode
and light
source (for
reference)

graticule

Fig. 8 A typical photometer for use with a PbS cell on a Cassegrain telescope.

optical axis and to this end it is mounted on a movable stage controlled by two screws.

After reflecting off the beamsplitter, the infrared radiation must be chopped. The photometer illustrated in Fig 8 is equipped with a motor-driven rotating chopper. The mounting of the motor must be carefully made to prevent transmitting vibration to the detector thereby causing microphonic noise.

THE DEWAR

To cool the detector to the desired temperature involves the use of cryogenic liquids which boil away rapidly in air and must be kept in dewars. In a dewar, as in a thermos flask, the container for the coolant is isolated from external heat sources by a vacuum. The detector is mounted on a metal block in good thermal contact with the coolant; it is therefore separated from the telescope by a vacuum space. The dewar is fitted with a small transparent *window* which can withstand the vacuum behind it and through which the detector 'sees'. Typical window materials for infrared use are sapphire, potassium bromide and various plastics.

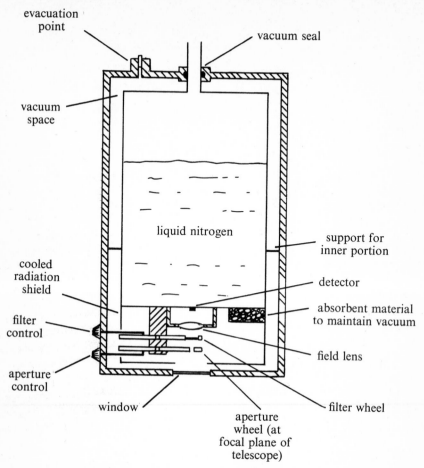

Fig. 9 A liquid nitrogen dewar to contain a PbS detector.

Most infrared detectors function poorly if much radiation is falling on them. Thermal noise, too, is introduced if the temperature of the material seen by the detector is not controlled. The incident radiation is minimised by cooling the surroundings of the detector: by Stefan's law the radiation is drastically reduced if the detector sees only material at liquid nitrogen temperatures. A series of cooled baffles is used to define the incoming beam so that the only material seen by the detector which is not cooled is the optical train of the telescope and photometer. For the same reason it is normal practice to cool the filters which define the waveband used and the apertures which define the beam diameter. These are mounted on the cold inner container of the dewar and are controlled from outside by linkages with a minimum of thermal contact. The aperture is at the focal plane of the telescope and is imaged onto the detector by a *field lens* of a material similar to that of the window.

Fig 9 shows a typical liquid nitrogen dewar for a PbS cell. If liquid helium is required the dewar becomes considerably more complex and may involve the use of a liquid-nitrogen jacket between the helium container and the external surfaces. To mention but one complication, the boiling of liquid helium in a dewar may produce excessive microphonic noise at the detector; one cure for this is to put a Brillo pad in the liquid-helium jacket!

When mounted onto the photometer it is necessary to align the optical axis of the dewar with that of the telescope and photometer. If this is not done the detector, viewing the outside world through the field lens, aperture and cooled baffles, may see only part of the main mirror and thus receive less than the maximum signal. More seriously if it is not so aligned it will receive the unwanted thermal radiation from the walls of the telescope and the primary mirror cell.

Since perfect alignment cannot normally be achieved in a rigid system, the dewar is usually attached to the photometer by a series of adjusting screws with which it can be tilted and moved laterally to the optimum setting. This is done empirically when viewing a bright star. The relevant adjusting screws may be fitted to either the photometer or the dewar. Once set up, the dewar must be tightly clamped to prevent its moving when the telescope is tipped to low altitudes.

Dewars used with Low bolometers are dangerous things. At the end of the night's observing if the dewar is to be removed from the telescope, the natural sequence of events is to disconnect the pump hose, loosen the clamps which hold the dewar to the photometer and remove it. If this sequence is followed the observer will find himself with a potential bomb in his hands.

The dewar has been evacuated. When the pump is disconnected air rushes in to equalise the pressure. But the air, and especially the water vapour it carries, freezes at liquid-helium temperatures. A plug of solid air and ice will form in the narrow neck at the top of the dewar. When this happens the helium container is closed off, and as the helium inside slowly boils it builds up pressure. After a period typically of the order of half an hour to two hours the pressure is so high that the dewar explodes. If the dewar is well constructed this explosion could wreck an observatory and easily kill the observer. There are a number of ways of ensuring that a plug does not form; the best is to blow helium gas through the pump and into the dewar to bring the pressure back up to atmospheric or a little above. The pump may then be safely disconnected.

No infrared astronomer is worth his salt until he has had the job of freeing a plug from the neck of a dewar before it explodes.

Astronomers are inclined to take very much for granted the electronic gadgetry they find themselves using more and more as they ply their trade. To some the 'black box' which lies somewhere between the detector and the chart recorder is more akin to witchcraft than to science. The demands that modern astronomy makes on electronics are ever increasing, and it is a tribute to exponents of the latter field that they are always able to come up with an answer to these demands.

When PbS and Ge:Ga detectors were first employed, the amplifiers necessary to make something measurable from the tiny changes of resistance they underwent were near the available limits. This is no longer the case, and a good range of suitable amplifiers can now be bought off the shelf.

Very briefly, the usual arrangement is as follows: a small voltage is applied across the detector in series with a resistor of similar impedence. Changes in the resistance of the detector therefore produce corresponding changes in the voltage across it, and when a chopped signal falls on the detector, a ripple is generated in the

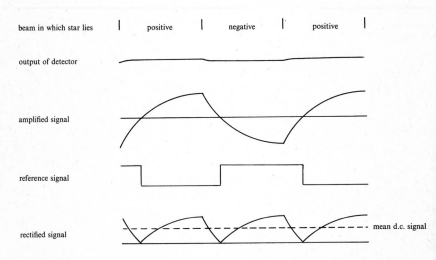

Fig. 10   Stages in the electronic processing of a signal

34

output voltage. The amplitude of this ripple is so small that if the signal were carried down long cables from the telescope to an amplifier on the dome floor, it would not survive the journey. A preamplifier is mounted on the dewar, and this typically amplifies the voltage ripple a thousandfold.

The main amplifier, having raised the voltage ripple to a suitable size, must then rectify it. Because of the slow response of the detector the ripple is not a square wave. When the source is presented to the detector the voltage rises slowly. Similarly when the empty beam is presented the voltage falls slowly. The signal from the detector and the amplified signal are shown in Fig 10. The amplifier cannot know when the rectification is to take place; it must be told. A *reference* square wave is introduced so that when the reference signal is on the amplified signal is treated as positive; when the reference is off the sign of the amplified signal is reversed to give the rectified signal shown in Fig 10. When this is smoothed it becomes a DC voltage proportional to the amplitude of the original ripple. This voltage may be fed into a chart recorder.

The output of the amplifier is also fed through a voltage-to-frequency converter and a frequency counter. This combination produces a digital output suitable for integration; after integration this may be printed out or punched on paper tape or on cards for processing by computer. The normal integration time is 10 seconds in each beam, and a run of observations of an object may involve as many as several hundred of these pairs of integrations.

In order to have exactly the right frequency, the reference square wave must be generated by the mechanism that produces the chopping. In Fig 8 the reference is produced by the rotation of the segmented chopper intercepting a light beam. A similar system is used when a vibrating mirror does the chopping, but if this mirror is electrically driven, as with the wobbling secondary, the electrical impulses which generate the chopping can be used directly as the reference signal to the amplifier.

REDUCTION OF THE DATA

The chart record is normally used only as a check that all is well. If, for example, birds or moths are flying through the beam, spikes will appear on the chart which may have a significant but unnoticed effect on the integrations. Or if seeing is a serious problem the chart record for a bright star will be ragged whereas the integrations, by averaging the effect over 10 seconds, may be quite uniform. Fig 11

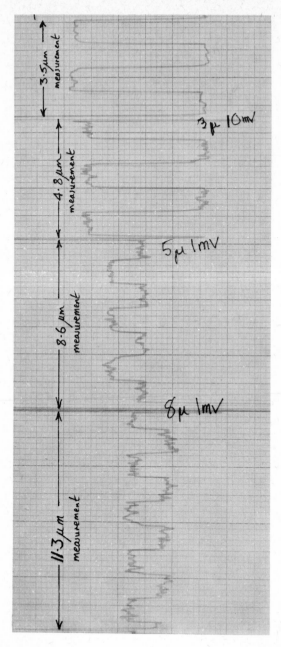

Fig. 11   A section of chart record. The measurement at 3·5 μm was made on a scale ten times smaller than that of the other colours.

is a section of a typical chart during the measurement of a bright star (actually Vega). Each square-topped section results from the source being in one or other beam and thus represents the interval between the nodding of the telescope.

It is the integrations which are finally used for quantitative measurements. The data take the form of a series of numbers which can be treated in a statistical manner. Indeed, the reason that integration times as short as 10 seconds are used even for faint sources which require, say, one hour of integration for a good measurement is that a statistical analysis can be made. If the noise behaves in a perfectly statistical way, the accuracy with which the true signal from the star is determined increases roughly as the square root of the time spent observing it. The series of numbers is averaged and the mean ideally represents the best estimate of the true signal. The median may be a better estimate in fact.

The data must be treated to a thorough statistical analysis. The standard deviation of the mean, σ, is used as an estimate of the noise, and the usual policy is to believe in any signal which has been integrated to a mean greater than 3 σ. Such a signal would normally arise by chance once in every 740 measurements. Some observers seeking shortcuts use 2 σ as their criterion, but this is dangerous, especially at long wavelengths, because the sky can sometimes generate spurious signals at the 2 or even the 3 σ level. When 10 μm photometry was being pioneered, therefore, a number of erroneous measurements appeared in print. Most of these have since been retracted or refuted. Even now in the 1970s a few observers make the same mistakes by being hasty and incautious in taking their observations. It is perhaps relevant to make the aside that estimates of the errors in and reliability of the observations are published in a smaller proportion of astronomical papers than of those of many other disciplines of science.

The raw integrations mean little in themselves: the sources observed must be compared with standard stars to determine their relative luminance. Infrared astronomy, leaning heavily on the optical, deals in stellar magnitudes, and on each night that data is taken a number of standard stars whose magnitudes are known must also be measured with the same equipment. If accurate data are required, about one eighth of the observing time must be spent in measuring standards. The standard stars are also used to determine how the atmospheric extinction varies with the altitude of the source. The correction necessitated by extinction can be quite large: at 20 μm a source may appear half as bright at 40° altitude as at the zenith. In the PbS region and the 8–14 μm window, extinction corrections are rarely more than a tenth of a magnitude. Not only is the extinction high at 20 μm, but on most nights it varies, usually being smaller towards dawn as the water vapour settles out.

The relative magnitudes of the standard stars must be determined accurately; this is done by frequent measurement. A zero point to the system of magnitudes at infrared wavelengths must be adopted, and Johnson's definition is used: namely that a normal star of spectral type A0 has the same numerical magnitude at all wavelengths. Vega (α Lyrae) is the infrared astronomer's principal standard, for it is a zero-magnitude A0 star. At 20 μm, however, Vega is too faint to be accurately measured, and the calibration most commonly employed is that in which the stars α Tau, α Boo and β Peg are given the same magnitude as at 10 μm. This is equivalent to Johnson's definition since these stars, like A0 stars, should be on the Rayleigh–Jeans tail of the Planck function where the shape of the energy distribution is the same for each.

Conversion to stellar magnitudes is not the end of the data reduction. If we wish to compare the measurements to black-body functions, monochromatic energy fluxes are needed. The conversion from magnitude to flux is deceptively simple: for each waveband a calibration exists giving the flux from a zero-magnitude star. Thus at 3·5 μm we receive at the earth $2·5 \times 10^{-14}$ Watts per square centimetre from a zero-magnitude star. From a fifth-magnitude star the figure must be $2·5 \times 10^{-16}$ W cm^{-2}, and so on.

The deception lies in the calibration itself which was not handed down as a gospel truth but which had to be derived. For every waveband used by infrared astronomers, hours of thought and measurement have been expended in deducing the best calibration. Even now the calibrations employed are little more than inspired guesses. So long as we must observe through the atmosphere we cannot simply measure the flux from a zero-magnitude star because we cannot estimate the atmospheric extinction or the transmission of the telescope and photometer. The ruses adopted to overcome this problem include measuring stars whose fluxes one might hope to predict using sophisticated models; the Sun is one such star. Solar-system bodies believed to be in equilibrium with sunlight can also be used, for their temperature and hence their flux can be predicted. Most infrared observers derive their own calibrations, and in the early days these differed by as much as 50 per cent. Subsequent revision and refinement have reduced the discrepancies to under 10 per cent, and this might be taken to represent the probable uncertainty in any one calibration. Perhaps the best calibrations at the longer wavelengths have been derived by comparing standard stars to the planet Mars at a time when a Mariner space probe was making infrared observations of the planet which it compared to laboratory standards carried on board.[3]

It is appropriate to end this chapter in a cautionary way by stressing the low accuracy of infrared photometry. Errors in the measurement of a star are compounded from the following sources:

statistical error of the measurement;

the effects of seeing;

the difficulty of centring optically faint sources in the beam and of guiding on them;

variations in the transmission of the sky with time and with position;

variations in the sensitivity of the detector and/or electronics;

uncertainties in the determination of the atmospheric extinction;

the effect of the above factors on the standard stars measured on the same night;

errors in the assumed magnitudes of the standard stars arising from the above factors operating on previous nights;

uncertainties in the calibration from magnitude to flux.

A handy rule of thumb – and it is no more than that – is that the best attainable accuracy in a single night's observing, expressed as a percentage or in hundredths of a magnitude, numerically equals the wavelength in microns. For J this is optimistic; at 350 μm it is probably pessimistic.

3
THE FORMATIVE YEARS

Lead-sulphide cells were put to use by astronomers long before doped germanium detectors. This was entirely natural, partly because the PbS region lies closer to the visible, but mostly because the techniques involved are considerably less complicated. The use of lead sulphide as an astronomical detector was pioneered independently in the USA by Whitford, who made few measurements, and in England by Felgett, then a research student in Cambridge. Much of Felgett's time was spent grappling with the new techniques, and it was only in the closing months of his studentship that he began to make measurements of some of the brighter stars.[1] His results, just like those of the earlier workers, showed nothing spectacular or unanticipated. Infrared astronomy was, it then seemed, doomed to a position subservient to the optical, and in England most research was shelved. The initiative went to the USA, and there it has remained since.

Johnson is an indefatigable worker. He initiated the photometric systems still most popular in optical photometry: his UBVRI bands have been mentioned in Chapter 2. During the 1950s and early 1960s he and his co-workers at the University of Arizona set about making very accurate measurements of several thousand bright stars in order to delineate the nature of normal stellar continua and of interstellar reddening. Johnson was quick to realise the relevance of the PbS region to his investigations of reddening. The grains of dust which lie in the interstellar gas clouds like some galactic smog form a significant barrier to stellar radiation in the blue and violet. Red and infrared photons, having longer wavelengths, stride past these grains virtually unimpeded. It is the effects of this on the energy distribution of stars which is called reddening. The dependence of the extinction on wavelength conveys information on the nature of the dust grains responsible, hence Johnson's interest in measuring reddened stars. To determine the wavelength dependence – the *reddening law* – involves knowing how bright each reddened star would be were the dust not present. No better method presents itself than to measure the star in the infrared where the reddening has the least effect.

From the Catalina Mountains outside Tucson, Johnson's team measured several thousand stars in the newly instituted *J*, *K* and *L* wavebands.[2] Many of the data were used to delineate the behaviour of normal stars at these wavelengths, and we owe much to Johnson's patience in carrying through what must have been a tedious task. Without this information on normal stars we should be considerably hampered in finding and understanding stars with abnormal behaviour. The particular abnormality Johnson sought was large values of the interstellar reddening, and by comparing reddened stars with normal stars on his list he derived the interstellar reddening laws he sought.[3] The curves depicting these laws often showed peculiar kinks in the infrared, and it now seems likely that these arise not from irregularities in the reddening laws but from abnormal infrared emission by the underlying stars. At the time, however, so few of the data in the Catalina photometry hinted at abnormal infrared behaviour that Johnson was led to believe that it was the absorbing dust which was responsible. In fact Johnson drew attention to only one peculiarity of the stars in his sample, namely that the B stars which showed bright emission lines of hydrogen and other elements in addition to the usual spectral absorption features – the Be stars – are systematically brighter in the infrared than those B stars without emission lines. This phenomenon is explained in Chapter 6.

At much the same time that Johnson was establishing the Catalina photometry, a group at the California Institute of Technology (CalTech) was making an equally significant contribution. Neugebauer and Leighton, both professors in the physics department, undertook to survey the entire northern sky at 2·2 μm. This was no small undertaking, for the sky is a very big place, and their survey took approximately six years to complete. The survey was made against the advice of the astronomers of CalTech who maintained that it would reveal nothing but a few red stars.

First Neugebauer and Leighton needed a telescope, for no observatory in the world could spare sufficient observing time on a large enough instrument. They designed and built their own, an inexpensive instrument made mostly from war-surplus materials. It employed several novel features of which the mirror is the most famous. The aperture is 62 inches and the parabola was figured not in an elaborate optical workshop but on a lathe. Glass, of course, cannot be turned on a lathe, so the mirror was made from aluminium and this was hollowed out in an engineering workshop to the desired paraboloid. The surface which resulted from this treatment was far from good. Neugebauer and Leighton therefore made use of the simple principle that if a bowl of liquid is rotated about a vertical axis its surface assumes a paraboloidal contour. They filled the aluminium dish with epoxy resin and spun it at constant speed until the epoxy set firmly into the desired shape. This technique cannot be recommended for the production of optical telescopes, for even this paraboloid leaves much to be desired: the image of a star at the focal plane of the mirror is some 4' arc across. For the purposes of the infrared survey, however, 4' arc was quite acceptable. In order to survey in a reasonable time the 30,000 square degrees of sky accessible from California, the beam employed must be a sizeable fraction of a degree across.

The 62-inch infrared telescope was used at prime focus, the dewar being mounted on struts in front of the mirror without an intermediate photometer. Chopping was accomplished by vibrating the primary mirror back and forth, something which a glass mirror could not withstand. The dewar combined a 4×2 array of PbS cells and a detector working at 0.84μm whose purpose was simply to bridge the rather large gap between the visible and 2.2μm.

The year 1969 came round before the results of this survey were published,[4] although a few of the more interesting sources were singled out for early release.[5,6] Even then not all the data were published. Neugebauer and Leighton decided to produce what they believed to be a complete survey. They expected to have detected all sources brighter than magnitude 3 at 2.2μm, but to have possibly missed some a little fainter than this limit. The infrared catalogue (IRC) contains 5,612 sources north of declination $-33°$ to a limit of $3^m.0$ at K. With the exception of two stars detected but accidentally omitted, we know of no 2.2μm sources brighter than this limit which were not included. On the other hand, although stars as faint as $4^m.5$ at K were picked up, there are several in the range 3.0 to 3.5 which escaped detection. The number of celestial sources brighter than magnitude 3 at K is thus roughly equal to the number of visible stars in the sky, ie those brighter than magnitude 6 at V. This leads to a handy rule of thumb: a star of apparent magnitude m_K at K is about as commonplace as one of magnitude m_{K+3} at V. Thus, the Sun apart, the brightest star in the sky at 2.2μm, Betelgeuse, is about magnitude -4.5 which compares with -1.5 for Sirius in the visible. The corresponding magnitude difference at 10μm is about 5.

The astronomers who claimed that nothing but a few red stars would be detected by Neugebauer and Leighton have been proved unduly pessimistic. The IRC contains a wealth of objects radiating unexpectedly strongly at 2·2 μm, objects which would have long remained unknown, and whilst it is true that a very large proportion of the stars detected are cool, red, late-type objects, many of these are emitting considerably more radiation than was predicted before the 2 μm survey was made. The influence of the IRC on the subsequent development of infrared astronomy has been marked, and in the observational chapters which follow frequent reference to it will be made. It is therefore useful to explain now the numbering system employed by Neugebauer and Leighton.

The first publication, by Neugebauer, Martz and Leighton, contained two of the most extreme infrared objects from the survey.[5] These have subsequently been known as NML Cygni and NML Tauri, after the constellations in which they lie. Both are visible objects but are very faint at optical wavelengths. NML Cygni is about magnitude 17. NML Tauri is a variable star at both optical and infrared wavelengths and has since been designated IK Tau. A further fourteen interesting sources were later released[6] and were given the designations CIT 1 to CIT 14. The initials merely stand for California Institute of Technology. Several of the CIT sources were known variable stars and already had names in the *General Catalogue of Variable Stars*; infrared astronomers, however, invariably use the CIT designations in preference. Two of these, CIT 1 and CIT 10, are invisible on even the Palomar Observatory sky-survey prints.

The infrared catalogue, being a complete list, contains the NML and CIT sources, but like the rest they are given IRC numbers. The survey was divided into declination bands 10° wide and numbered along each band consecutively in R.A. A designation IRC +20 435 therefore means that the source lies in the range +15° to +25° declination. The last three digits are its number in the band, and since there are about 500 sources in each band (except those near the pole), number 435 is at about 20 hours R.A.

An essentially similar survey of the southern sky was begun by Stephen Price. He chose the Mount John Observatory in New Zealand as his base, but acting essentially alone he found the task too onerous to complete. A partial survey was published containing 414 sources; many of these were bright stars.[7] Very little attention has been paid to Price's survey, largely because of the dearth of infrared facilities in the southern hemisphere.

During this period and subsequently, the limited information that has come out of the Soviet Union shows that the Russians were using PbS and other infrared detectors at much the same time as the Americans. Very little has since emerged, partly because Russia lacks good infrared sites, and it would seem that Soviet interest in the infrared has subsequently been rather slight.

THE 10 μm REGION

By the time that useful infrared observations in the 5–10–20 μm (doped Ge) region were being made, PbS observations had reached a high level of sophistication. Ge:Cu cells were employed by Wildey and Murray in the early 1960s to measure a few of the brightest stars.[8] The fluxes they measured from some stars have since been shown to be grossly inaccurate, two particular cases in point being α Leo and γ Ori.[9] A similar error was made by Mitchell, another pioneer of this period, in respect of the star ε Aur.[10] This observation has led to some misunderstanding in recent theoretical papers on the nature of the mysterious companion to ε Aur.

It would be churlish to stress the shortcomings of the early results. Photometry in the 10 μm region is a difficult business at the best of times and with the insensitive detectors then in use it was doubly so. The only mistake made by the first observers was to believe their results – an entirely understandable reaction – and it is only recently that we have reached sufficient understanding of the 10 μm region to know when the signals apparently coming from a star are in fact spurious in nature. Little reliance can be placed on the actual measurements reported during the first five years of photometry in the doped Ge region – ie up to about 1969 – although the fundamental results and principles deduced therefrom are generally sound.

The development of good 10 μm systems was a slow process, and credit for much of the work undoubtedly belongs to Frank Low. As a physicist, Low's contribution in inventing the Ge:Ga bolometer was already significant. He foresaw, however, that this detector could revolutionise infrared astronomy, and consequently took up a post at the University of Arizona where he set about establishing the 10 μm region as a respectable branch of astronomy. Had Low not changed his allegiance in this way it is likely that the taking of useful astronomical measurements in the 10 μm region would have been delayed several years. Nor was Low's pioneering spirit exhausted by this task: he has subsequently pressed on to longer wavelengths, publishing the first 35 and 350 μm observations in 1973,[11,12] and sampling the 100 μm region by taking telescopes above most of the earth's atmosphere in high-flying jet aircraft.

It is bad for science if only one man is capable of making a certain type of observation. The temptation is great to seek the glory of discovery and shun the routine but necessary repetitive work involved in accurate observation. Rarely is that one man motivated to retread old ground, finding, admitting and correcting the mistakes he has inevitably made. Therefore, once Low had shown the whole business to be possible, others tried to emulate him. Only one group at first succeeded.

The man largely responsible was Ed Ney, a physics professor at the University of Minnesota. Subsequently, Nick Woolf was appointed to a vacant chair at the university and became head of the infrared astronomy group. Ney and Woolf set up a conventional 30-inch telescope outside Minneapolis and, independently of Low (although, of course, relying on the knowledge that he had succeeded) solved the numerous problems associated with the manufacture of bolometers, their use at a telescope and the interpretation of the results. By 1968 the Minnesota system was functioning.

Low eventually founded a private company selling Ge:Ga bolometers and liquid helium dewars, and thus made available to the astronomical community at large the capability of infrared photometry. PbS photometers spread rapidly too, so that there are now many more astronomers and a great number of observatories practising this science. Too many, in fact, to list all of them. But a few are pre-eminent in the field, and these are mentioned now.

Mount Lemmon Observatory

The Minnesota site is a poor one. Although having the great advantage of convenience, being but an hour's drive from the university, the climatic conditions are abysmal. The altitude is low and in summer the humidity is high, resulting in a murky and unstable 10 μm atmospheric window. In winter the temperatures hover around $-25\ ^\circ$C and the water-vapour content of the air is very low, but only a few clear nights per month can be expected. The Minnesota team, together with a group at the University of California and with support from the Science Research Council of Great Britain, sought a better site on which to place a 60-inch telescope dedicated to infrared photometry. They chose Mount Lemmon.

The choice was a logical one. The desert of southern Arizona

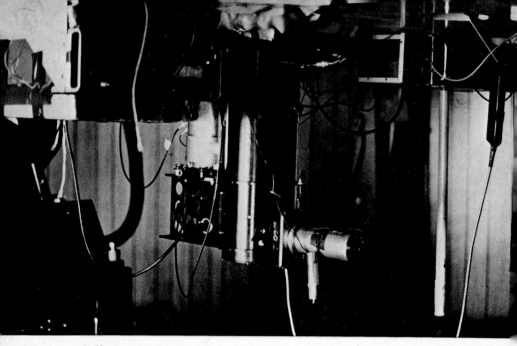

1 Photometer (black) and dewar (silvery) for 10-micron work mounted at the Cassegrain focus of the 60-inch Mount Lemmon infrared telescope. The amplifier and integrator are also mounted on the telescope.

boasts the lowest humidity and the highest frequency of clear nights in the mainland USA. Tucson is a major centre of population with all the amenities infrared astronomers desire, in particular supplies of cryogenic liquids. And within easy distance of Tucson are an assortment of knobby purple hills rising into even drier air, like the spoil heaps of an abandoned gold mine. Of these the highest is Mount Lemmon, at over 9,000 feet the crowning summit of the Catalina Mountains. Finally, at the very time the Minnesota/California team were seeking a site, an air-force base on the summit of Mount Lemmon was vacated.

A cheap telescope was built – a 60-inch, metal mirror, Cassegrain focus telescope designed solely for infrared work. It incoporates such features as a well behind the mounting in which the vacuum pump can be stored out of the way, and a pumping hose built onto the telescope instead of the usual arrangement where a hose hangs limply from the dewar and may cause vibrations leading to microphonic noise. The complexities of the Mount Lemmon telescope may be seen in the plate 1. A glass mirror giving better images and a wobbling secondary were installed in 1974.

Mount Lemmon has developed into a major infrared site, for in addition to the Minnesota/California telescope the instruments used by Low and Johnson have been moved from lower on the Catalinas to the summit site, and a further infrared telescope is operated there by the National Aeronautics and Space Administration (NASA).

More recently infrared astronomers have made partial takeovers of the telescopes at Kitt Peak National Observatory, to the west of Tucson, and of Mount Hamilton, the Smithsonian Astrophysical Observatory facility at the southern edge of the Tucson basin. This region is the major infrared centre of the mainland USA; while not the best of infrared sites it is the most accessible of the good sites.

Mauna Kea

The site in the northern hemisphere which, at the time of writing, is the best known for infrared observing is also on American soil, in the Hawaiian Islands. Mauna Kea is volcanic and the highest peak on the island of Hawaii, the largest member of the Hawaiian archipelago. It rises to almost 14,000 feet above sea level, yet although the highest, Mauna Kea is not the principal mountain of Hawaii. It is a steep cinder cone, a pimple on the flanks of Mauna Loa – the 'long mountain' of the Polynesians and the largest mountain in the world. From the ocean floor to the steaming summit caldera Mauna Loa measures a full 30,000 feet, and every inch of this was built from lava flows extruded from its active vents.

Although Mauna Loa is still active, there are good reasons for believing that its subsidiary vent, Mauna Kea, is extinct. There are, of course, even better reasons for hoping this is so. Certainly there has been no volcanic activity since the Ice Age, for a small cinder moraine lies just below the summit. Mauna Kea is an utterly desolate spot. This mound of cinders is barren of all forms of life (except astronomers) on its upper slopes. Even the living quarters at Hale Pohaku, 4,500 feet below the summit, lie above the limit of vegetation. There are no cliffs, no streams, no birds; merely an endless amorphous sea of brown and purple cinders. There are also no clouds – or at least very few of them – and at 14,000 feet there is very little atmosphere. It is therefore not surprising that Mauna Kea is excellent for infrared astronomy. So good a site is it that nearly all ground-based observations beyond 25 μm have been made from its summit.

The first major telescope to be used on Mauna Kea for infrared work was the 88-inch reflector of the University of Hawaii which for many centuries ruled the roost there. More recently the French and Canadians, collaborating on the construction of a major optical telescope, have chosen Mauna Kea as their site, and it now seems likely that this telescope will be extensively used for infrared work. Britain proposes to build a 3·8-metre infrared telescope on Mauna Kea and may also site a major new observatory there. Finally NASA will probably site a 3-metre infrared telescope there.

This, then, promises to be the infrared centre of the astronomical world and anybody wishing to be an active infrared astronomer must therefore be prepared to work there. An observing run on Hawaii may sound like a free ticket to a Polynesian paradise, but the harsh realities of working on Mauna Kea will have little appeal to those who contemplate lazy afternoons on black sand beaches, strolls through orchid fields or rough scrambles into the gaping volcanic craters of the National Park. The altitude to which Everest climbers descend for recuperation may be 14,000 feet, but to ordinary mortals who dwell in the heavy air of sea level it is no balm. Some will quite simply not be able to work there. Others will find their minds shrinking into tangled oblivion as oxygen deficiency gnaws at their bloodstreams. Some will be quite unaffected. Short spells may be tolerable; the sort of long-term observing runs that are necessary to the economic use of observatories at great distances from home base could prove disastrous to all but the most versatile of metabolisms.

Nor is the altitude all. To spend three weeks on a sterile cinder heap is nobody's idea of fun. Better the Antarctic wastes where at least the occasional penguin or skua might appear. The mental fatigue of living in an absolutely uniform world may also take its toll of future infrared astronomers.

Tenerife

When Piazzi Smyth chose the volcanic island of Tenerife from which to make his infrared detection of the moon he argued that it was a high dry site where one would expect good atmospheric transparency. More detailed meteorological records now exist which amply confirm Smyth's assertion. Tenerife, like some of the other islands in the Canary archipelago, is a superb astronomical site ideally suited to infrared observing. It was therefore chosen as the site for Britain's first infrared telescope, a 60-inch reflector built by Ring's group at Imperial College, London University. In many respects Tenerife and Mount Lemmon are similar. The observers operate the telescope themselves and once 'on the mountain' are essentially on their own. If something goes radically wrong it may mean the end of the observing run. Unlike Mount Lemmon, however, there really are no facilities in the valley on which to call. The observers' job is therefore that much more difficult. By way of recompense, instead of having to do their own cooking, laundry etc, as at Mount Lemmon, the Tenerife observers stay at a resort hotel a little further up the mountain.

More important differences exist between the two establishments. For one, the mirror of the Imperial College telescope is of glass, in contrast to the metal mirror originally installed on Mount Lemmon. The Mount Lemmon optics gave extremely bad images: at best not more than 80 per cent of the radiation from a star entered the 1 mm aperture usually employed. Often during the night as the temperature, and hence the figure of the mirror, changed this percentage could be much reduced. The glass primary at Tenerife was little more expensive, being made from an unusually thin mirror blank, and the images it produces are of order 1·5″ arc across. Not only is the photometry more reliable, but the limiting magnitude of the telescope is a good deal better.

At Tenerife, however, there is no standard photometer and dewar, and observers must take their own equipment. This restricts the users of the telescope to astronomers who have constructed infrared equipment, and results in unnecessary duplication of facilities amongst the various UK groups.

GETTING ABOVE THE ATMOSPHERE

Between 40 and 330 μm the atmosphere is like a dense fog. On a clear day you can see the sun and full moon, and not much besides. To make any useful observations through this waveband – in what is generally termed the 100 μm region – the detector must be carried to a greater height even than the summit of Mauna Kea. Three techniques of accomplishing this have so far been employed: aircraft, balloon and rocket. Each has both merits and demerits.

Aircraft can be used at 60,000 feet, at which height the atmosphere is still by no means clear. Their great advantage is that the observer is present to control the telescope, to point it where he wants, to integrate for as long as is necessary, etc. Balloons fly higher and do not carry human passengers. An elaborate system of remote control is needed, and the problems of stabilising the balloon gondola and of pointing the telescope are great. There is also the possibility of loss of the equipment if the upper winds have been poorly estimated, or of damage to it during its descent.

Rocket flights permit observation from above the last vestiges of atmosphere. The duration of a flight is short and pointing is particularly difficult. Observations are frequently made as the rocket spins, so that scanning of the sky is possible. Rockets are therefore used extensively for survey work, whereas balloons and aircraft are generally more suited to detailed observations of known sources. The first 100 μm survey was that of Hoffmann, Frederic and Emery which recorded over 70 sources.[13]

In 1974 the United States Air Force published an infrared sky survey made by rocket-borne detectors.[14] The full details of the survey, including the limiting magnitude, are classified information and cannot be given here. The catalogue contains observations at 4, 11 and 20 μm of 2,066 celestial sources. In addition a large number of solar-system objects were recorded.

The publication of the USAF survey, and indeed its early circulation to a select number of infrared astronomers, promoted a flurry of observations of the objects it contains. Many of the entries are IRC sources and were therefore already under investigation; the new sources excited the greatest interest. Observers with ground-based telescopes devoted many hours to searching for the 11 and 20 μm sources at the approximate coordinates given in the USAF survey. These searches proved only about 30 per cent successful. The problem, it now turns out, is not that the sources are spurious, nor that their position was poorly determined by the rocket survey. Rather, the 10–20 μm sky contains a large number of extended sources – infrared nebulae – which are bigger than the beams with which terrestrial telescopes are equipped.

The longer the wavelength at which we observe, the larger the proportion of extended sources we find in the sky. As we progress from 1 μm to 10 μm, we find ordinary stars becoming less prominent; instead the nebulae – planetary nebulae, gaseous nebulae, galaxies – take over. In describing the infrared sky we will be making mention of every type of visible object and more besides. No aspect of optical astronomy has proved immune to the impact of the last few years' infrared work, and for most objects the results that have emerged have been far from those expected.

4
THE SOLAR SYSTEM

At optical wavelengths it is often convenient to divide the solar system into two types of objects according to whether they emit light on their own behalf (the sun) or are made visible by reflecting sunlight (the rest). At wavelengths beyond about 5 μm no such distinction can be made: all objects are radiating thermally. The interpretation of this thermal radiation varies considerably from object to object, however, and it is essentially necessary to deal with each as an individual.

THE SUN
The sun is not a solid body but a mass of ionised gas. How far we see into this gas is a function of the wavelength we use. In the optical at about 0·5 μm the gas is fairly transparent and we see a long way in. At radio wavelengths the gas is very opaque: the sun appears much larger because its radio radiation originates further out. The infrared essentially spans these two domains, and by choosing the wavelength we can select how deep into the sun we observe.

Detailed models of the sun and its outer atmosphere predict the size and apparent temperature of the sun as a function of wavelength, and these are simple matters to check. Much of the solar infrared research has been devoted to determining the apparent temperature at all convenient wavelengths, and the agreement between theory and observation is heartening. The match does not seem so good concerning limb darkening. We are used to seeing the sun darker at the edges, but theory predicts the amount of limb darkening to decrease at infrared wavelengths until in the far infrared the limb should be brighter than the centre of the disc. The observations do not confirm this. Although limb darkening does vanish in the Ge:Ga part of the infrared, limb brightening is not observed at longer wavelengths. Minor modifications to models of the sun will be necessary.

There has been little work devoted to mapping the more transient features of the solar disc – sunspots, faculae etc. Many of these are not readily observable because the phenomena are less pronounced at higher levels of the solar photosphere. However, there is much of this sort of work to be done at 1·65 μm, for here the opacity of the sun is lower than in the visible, allowing us to observe even further in.

The infrared is also the domain of molecular transitions. Most gaseous molecules have absorption bands at infrared wavelengths, and small quantities of the more stable molecules should be present in the sun. The only difficulty in observing these lines is that few of the useful ones occur in terrestrial atmospheric windows. Observations from way above the atmosphere are needed.

THE MOON

The large temperature range of the lunar surface from day to night and in eclipse, mentioned in Chapter 1, was the first interesting result to emerge from infrared photometry. This will now be analysed in a little more detail since it introduces a theory relevant to all solar-system bodies without atmospheres. Let us first consider the behaviour of an object which remains in perfect equilibrium with the incident sunlight, and let us place this object on the moon and examine its temperature variation. During one synodic month ($29\frac{1}{2}$ earth days) this object will experience one protracted day in which it is illuminated by the sun for just half of the time. If we lay it flat on the moon's equator, the sun will pass directly overhead.

During the lunar night our hypothetical object receives no sunlight; because it absorbs nothing it cannot radiate and its temperature must be absolute zero. When the sun is in the sky the temperature of our object is such that the energy radiated, given by Stefan's law, equals the energy absorbed. This is a maximum at midday when the sun is at the zenith; at other times the solar radiation is reduced by foreshortening. Immediately at sunrise or sunset, sunlight hits the object at grazing incidence and very little is absorbed. As the sun rises, the energy absorbed by the object follows a cosine law. We can write:

$$\sigma T^4 = L \cos \phi \qquad\qquad A$$

where L is the solar constant (137 W m^{-2}) and ϕ is the angle between the sun and the zenith. If we write L in the form σT_0^4, we have:

$$T = T_0 (\cos \phi)^{\frac{1}{4}} \qquad\qquad B$$

and T_0 is seen to be the equilibrium temperature for vertical incidence of sunlight ($\phi = 0$). T_0 is, in fact, the maximum temperature experienced, and in this case it is also equal to the range in temperature

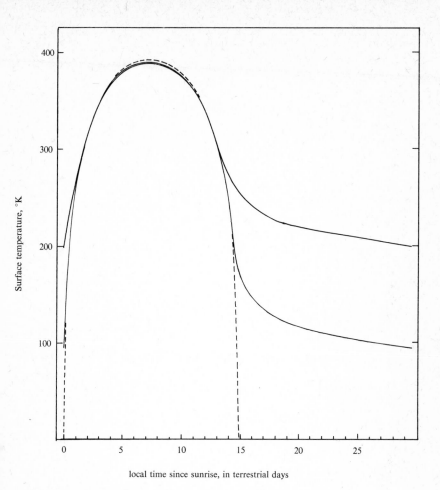

Fig. 12 Temperature curves throughout a lunation. The dotted curve is for a surface with no inward conduction of heat; the solid curves are for surfaces with thermal inertias of 20 and 1000, corresponding to solid rock and dust respectively.

from the middle of the lunar day to the depths of the lunar night. Fig 12 shows the variations of T during one complete lunar day. For the moon, T_0 is 394 °K.

The surface of a planet does not assume instantaneous equilibrium with incident radiation. Although struggling to do so it is always lagging behind. The reason is quite simply the existence of thermal conductivity. During the lunar day some of the absorbed energy flows inwards from the surface to warm the interior. The surface temperature is therefore reduced slightly. The heat stored below the surface flows outwards again to maintain the temperature above absolute zero throughout the night. The lowest temperature occurs just before dawn.

The form of the temperature curve for a planetary surface depends on just three factors. One of these, obviously, is the appropriate value of T_0, and another is the length of the day. For the moon both of these is fixed. The remaining parameter is usually called the thermal inertia, written γ, and is defined by:

$$\gamma = 1/\sqrt{\kappa \rho c}$$

where κ is the thermal conductivity, ρ is the density and c is the specific heat. For terrestrial rocks a typical value of γ is 20; a value near 1,000 is more appropriate to the dust of the lunar surface. Temperature curves for these two values of γ are included in Fig 10. Only finely powdered dust can attain a γ of 1,000: this is because there is very little contact between individual grains and κ is very small.

To an observer standing at the apparent centre of the moon's disc, the sun and earth are nearly at the zenith when the phase is full. If he were to walk an angle θ around the moon (in any direction), both would therefore have a zenith angle of θ. From equation B we can see that the temperature measured at a selenocentric angle θ from the disc centre at full moon should be $T_0(\cos \theta)^{\frac{1}{4}}$. This is Lambert's law. The empirical results derived by Pettit and Nicholson did not confirm this simplistic theory, however.[1] A better fit was given by a $(\cos \theta)^{\frac{1}{6}}$ dependence. The difference between the theoretical and empirical laws is most marked at large values of θ, ie towards the limbs. The moon emits more strongly near the limbs than Lambert's law predicts.

Lambert's law assumes the body concerned is a smooth surface. We know, however, that the surface of the moon is far from smooth. Irregularities occur on scales from microns to kilometres. When we look towards the limb of the moon we see primarily the facing slopes of mountain ranges and crater walls. These receive sunlight at a less-grazing incidence than would a level plane, so the temperature of the inclined surface is higher. When models are developed making allowance for the proportion of inclined surfaces seen at large values of θ, the departure from Lambert's law is readily explained.

Because the limbs appear brighter at full moon, the total flux from the moon is about 10 per cent greater than that from an equivalent Lambert surface. This is an important result for it has a bearing on observations of other solar-system bodies.

54

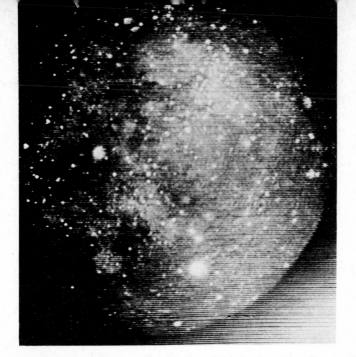

2. The moon at 10 μm during a total eclipse. The shadow moved across from the left; this portion of the disc has cooled more therefore, causing the smaller density of hot spots. The key identifies the brighter hot spots.

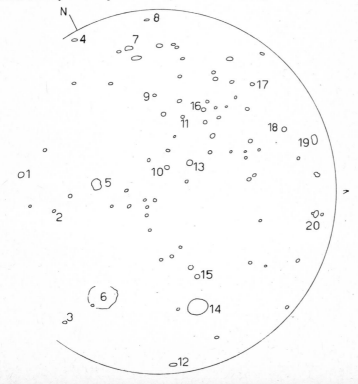

The value of γ appropriate to the cooling of the moon during eclipse is slightly higher than the figure of 1,000 found for the night. In order to explain the difference, Winter and Saari proposed that the thermal conductivity and specific heat of the lunar dust were not constant, but themselves depended on temperature.[2] By computing the behaviour of a surface having these properties, Winter and Saari produced an excellent match to all the observational data. Their paper was published in 1969, only a few months before man started to lay his heavy feet on the virgin soil, to stick thermometers into nook and cranny and to bring back for analysis precious ounces of the dust. Laboratory determinations of κ and c for lunar soil samples confirm that both these parameters are temperature-dependent in just the way Winter and Saari suggested.

Hot Spots

Perhaps the most exciting discovery of all lunar infrared research was made by Shorthill, Borough and Conley during the total eclipse of the moon of March 12–13, 1960.[3] Scanning the moon with a thermistor bolometer, they passed over the crater Tycho and found a signal from it more than treble that from the surrounding surface. Strong signals were also found from Aristarchus and Copernicus. It conveniently happened that in 1960 there were two total eclipses of the moon, and at the second of these Saari and Shorthill were able to add Proclus and Kepler to the list of what became known as *lunar thermal anomalies*, or, in the trade, *hot spots*.[4] (Richard Shorthill delighted in giving society lectures entitled 'Cold Facts on Hot Spots'.)

The next opportunity to study hot spots was the 1964 eclipse of the moon which Saari and Shorthill observed from the Kottamia observatory, Egypt. They made complete scans during totality at an effective wavelength of 11 μm. The scans recorded over 1,000 thermal anomalies. By displaying the scans on an oscilloscope screen and photographing the result, Saari, Shorthill and Deaton produced a picture of how the totally eclipsed moon would appear to a being with infrared-sensitive eyes.[5] This is reproduced in plate 2.

Hot spots can be traced throughout the lunar night, and most of them remain considerably warmer than their environs until dawn. At night, however, being cooler than in eclipse they are considerably more difficult to measure. No significant results emerged until Ge:Ga cells became available: their secrets were then prised from them with the 30-inch Minnesota telescope. Almost 300 were recorded in scans of the unlit portion of the moon.[6] The most prominent hot spots, both at eclipse and by night, are listed in Table 2.

If the night-time signal from the brightest hot spot, Tycho, were to be translated into a temperature, the value would be about 150 °K. This means that a 150 °K black body completely filling the beam would give the observed signal: 150 °K is its *brightness temperature*. However an area smaller than the beam and at a higher temperature than 150 °K could also give the observed signal. *The brightness temperature of an object is a lower limit to its actual temperature.* We can make a better estimate of the temperature by observing the object at different wavelengths and using Wien's law to determine its *colour temperature*. This was done for Tycho by Allen and Ney.[7] They found the colour temperature to be over 200 °K when the brightness temperature was 150 °K. About one tenth of the beam must be covered by 200 °K material to produce the observed signal. Similar results were found for other hot spots.

The colour temperature of the hot spots through the lunar night follows very closely the curve for $\gamma = 20$ in Fig 12. This suggests that thermal anomalies are caused by small areas of naked rock concentrated within certain craters. Inspection of high-resolution Orbiter and Apollo photographs enables us to identify the rock component as large boulders which lie in profusion on the slopes and floors of young craters. The lunar hot spots, therefore, are not flaming volcanoes or heated underground cities, but ordinary mundane lumps of rock. It is in fact interesting that by working in the infrared where the resolving power of our telescopes is rather low

TABLE 2

The Principal Lunar Thermal Anomalies

(Relative Intensity of Hot Spots over Background at 12 μm)

Tycho	100
Aristarchus	43
Copernicus	35
Mare Humorum (very large area)	25
Petavius B	16
Zuchius	14
Marius C	12
Messier A (Pickering)	12
Langrenus	11
Cauchy	10

On this scale the average background is 2 to 3

we can detect objects far too small to be resolved in the optical with the largest ground-based telescopes available.

The typical temperature of solar-system bodies is a few hundred degrees Kelvin. At wavelengths shorter than 5 μm the thermal radiation from them is extremely small and we record instead the sunlight they reflect, just as in the visible. For most planets, therefore, observations out to 3·5 μm tell us only the albedo at wavelengths in the terrestrial atmospheric windows: the real infrared begins at 5 μm.

Mercury is the exception to this rule. Being the closest to the sun it is also the hottest, attaining 700 °K. The thermal radiation from Mercury therefore peaks at about 5 μm and is considerably greater than reflected sunlight at 3·5 μm. Mercury is easy to study with a PbS cell, yet the only published observations have been made with the Minnesota germanium bolometers. Most optical observatories do not take kindly to astronomers pointing their telescopes within a few degrees of the sun.

Like the moon, Mercury is a good black body throughout the infrared. The emissivity is within a few per cent of unity in all the atmospheric windows. Unlike the moon it is too small for us to map its surface at infrared wavelengths. We do not therefore know whether it obeys Lambert's law. We can gain some idea, however, by examining the variations of the infrared signal with phase. It is a simple matter to predict the behaviour of a Lambert surface under these conditions. This experiment was performed by Tom Murdock for his PhD thesis at the University of Minnesota. Murdock found that Mercury is not Lambertian. He made the same observations of the moon throughout its phases and derived a similar curve, and from the comparison he could say that Mercury has a rough surface, as does the moon. It was not possible to say what form the roughness took; it required Pioneer 10 to show us that Mercury is cratered.

Considerable interest devolved upon measuring the temperature of the dark side of Mercury. This was a particularly difficult measurement to make because the illuminated portion of the planet emits so strongly at all infrared wavelengths that the tiniest portion completely swamps the signal from the dark side. There were only two possibilities: either use a very small beam to exclude all the illuminated crescent, or observe Mercury during its new phase. The former has not been seriously attempted and would in any case require a rather large telescope. But to observe Mercury at new is impossible because that phase occurs only during a transit of the sun! The best that can be done is to measure the planet as close as possible to the sun in the hope that the illuminated crescent makes a negligible contribution. Such an observation was reported by Murdock and Ney in 1970 who found a temperature of 111 °K for the dark side of Mercury.[8] Since they may have underestimated the effect of the tiny crescent we should regard this as an upper limit to the temperature. We thus deduce a lower limit to the value of γ, the thermal inertia. Murdock and Ney found a value of $\gamma \sim 700$ to fit their observations best. We can therefore state with confidence that the surface of Mercury is covered mostly with dust and not with solid rock. Although a slightly different value of γ is likely to result from the final analysis of the infrared measurements made by Pioneer 10 as it passed Mercury early in 1974, the conclusion will certainly not be affected. Whether Pioneer 10 recorded any hot spots remains to be seen.

VENUS

The Lambert model and its modifications which we used for Mercury and the moon is not applicable to Venus because of the thick atmosphere which girdles this planet. Sunlight does not filter to the surface of Venus: it is halted in its course and distributed throughout the opaque clouds much as is water dripping onto a sponge. Somewhere at the bottom of the atmosphere it must be hot – maybe as hot as the surface of Mercury. This much we have learnt from the various space capsules which have been despatched on their suicide missions to the planet's surface.

The clouds are opaque to infrared radiation too. It is not therefore possible to measure the surface temperature of Venus. The temperature we do deduce refers to a region somewhere in the thick of the clouds. A typical value at 10 μm is 220 °K, and this is much too low for a planet at Venus' distance from the sun; only an Eskimo would regard such a temperature as warm.

At 10 μm Venus has no phases. Even when only a thin crescent is illuminated, the infrared signal comes from the entire disc. At the depth into the clouds we sample at 10 μm there is no temperature change from day to night, and this implies that the atmosphere circulates freely. Early attempts at mapping Venus suggested that there are locally hotter areas. If there are they are almost certainly of a transient nature and may represent thinner sections of cloud allowing passage to infrared radiation from the deeper hotter layers.

The outer planets do not have sufficiently large phases for earth-based measurements to be made of the night-time temperature. And Mars is a little small to test its surface roughness by departures from Lambert's law. Thus there is not much one can hope to learn about its surface by earth-based infrared photometry. At all infrared wavelengths from 5 to 25 μm the energy distribution follows pretty closely a black body at a temperature of about 225 °K. Mars is a good source on which to check one's absolute calibration of the longer infrared wavelengths.

High-resolution maps have been made in sufficient detail to show that the dark areas are slightly warmer than the bright.[9] The most likely explanation of this is that the brighter portions of the planet reflect more sunlight and therefore absorb less. Hence they cannot attain so high a temperature as the darker patches.

Mariner 9 carried on board an infrared photometer which will be able to tell us something of the surface properties. Results for Mars itself have yet to be published, but observations of Phobos have been released, and from the temperature of the dark side it is possible to deduce that even this tiny world is covered with at least 1 millimetre of low-conductivity dust-like material.[10]

JUPITER

Jupiter, the largest planet, is also the most interesting at infrared wavelengths. At K it is a bit fainter than we would expect from its visual magnitude. At L Jupiter is almost invisible. At M it is much brighter than any simple model would predict. Only at 10 and 20 μm do we get reasonable results corresponding to 'surface' temperatures of about 140 °K.

The spectrum from 2·8 to 14 μm was described by Gillett, Low and Stein.[11] Throughout the 3·5 μm window they found the flux from Jupiter to be extremely small, and at some wavelengths they were unable to detect the planet at all. At L the signal is entirely contributed by reflected sunlight: Jupiter absorbs almost all the incident radiation ·and reflects very little. Gillett *et al* attribute this to the presence of methane in Jupiter's atmosphere. Methane has a strong molecular absorption band at 3·5 μm. Many of the other molecules in Jupiter's atmosphere could in principle be detected and their abundances measured in the infrared if only we weren't plagued by the absorptions of the earth's atmosphere which always seem to occur at crucial wavelengths. Molecular absorptions make Jupiter faint at 2·2 μm also.

So bright is Jupiter at 5 μm that it has been suggested that it radiates in this waveband alone more energy than it absorbs from the sun – in other words, that Jupiter has an internal heat source. When the long wavelength data are also included we find that Jupiter radiates nearly three times the energy it receives.[12] Contraction of the planet releasing gravitational energy is thought to be the source of this extra radiation.

The reason for the excessive 5 μm radiation is that Jupiter has an atmospheric window at that wavelength. At the longer wavelengths we see only the top of the atmosphere where the temperature is typically 140 °K. At 5 μm the atmosphere is fairly transparent and one sees deeper to layers of higher temperature. In some parts the 5 μm brightness temperature is as high as 225 °K. At 10 μm there is no temperature variation across the surface of the planet. Maps of the surface show considerable structure at 5 μm, however. It is particularly interesting to compare these maps with the visual appearance of the planet. The hotter and cooler sections of the planet can be identified with the dark and light belts which cross Jupiter respectively.[13] Plate 3 compares optical and 5 μm 'photographs' of the planet.

The equatorial regions, the NEB, SEB and EZ are hot. The dark belts at temperate latitudes (the NTB and STB in particular) are usually warm. The light zones are coolest. Dark blue and brown areas are usually the hottest of all. In the polar regions there are local hot spots less than 1″ arc in diameter.

It seems that the brightest portions, including the white spots and ovals, are high clouds floating some distance above the lower brown and blue areas. This result is confirmed by detailed optical studies of the behaviour of the white spots.

At 7·8 μm there is another strong methane band. At this wavelength Jupiter was recently found to have bright limbs.[14] The interpretation of limb brightening is that in one part of Jupiter's atmosphere there is a temperature inversion, so that one layer is warmer than that immediately below it. The structure of Jupiter's atmosphere is therefore seen to be rather complex, and it may be some time before it is fully understood.

A very curious result emerged from the work of Murray, Wildey and Westphal in the early days of 10 μm photometry.[15] They were using the Palomar 200-inch telescope to make maps of Jupiter, and in the course of scanning the planet they recorded a strong signal as

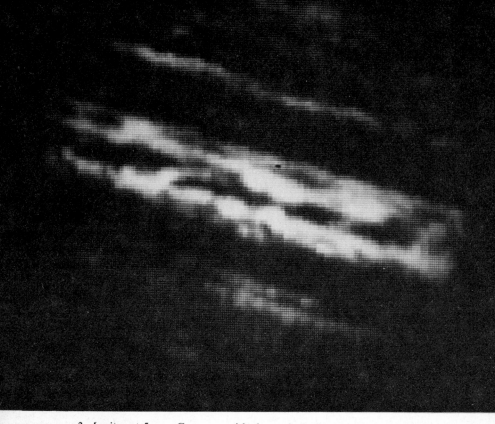

3. Jupiter at 5 μm. Compare with the optical photograph, taken a few minutes later.

the beam passed over the shadow of one of the Galilean satellites. The same phenomenon occurred on more than one occasion and with the shadows of different satellites. Why should the area shaded from the sun be so much hotter? This was a question to which no easy answer could be found. One might invent some mechanism whereby the upper layers of the atmosphere were dissipated allowing radiation from the lower and hotter regions to escape. But no such mechanism could operate in the short time scale involved when a satellite shadow sweeps across the planet.

These mysterious observations have not been explained. But perhaps they need not be. Since that early result no observers have found hot satellite shadows on Jupiter. The phenomenon, if real, is certainly transient or rare, and it is perhaps safer for the time being to attribute it to some spurious cause. Perhaps a bird flew through the beam at the critical moment. As I have said, the pioneers of 10 μm astronomy had a lot to learn about their subject.

The Galilean Satellites

The four principal satellites of Jupiter are not difficult to detect at 10 and 20 μm. Since all four seem very similar at optical wavelengths it might be supposed that their infrared properties are equally comparable. This is not the case: their infrared fluxes span a large range. At 10 μm satellite IV (Callisto) is very bright whereas I (Io) and II (Europa) require some integration for a good measurement.

Several factors contribute to the differences in infrared brightness, but the major two are the size and the optical albedo. IV is large and dark; I is small and bright. Although they appear similar in the optical, the larger and darker of the two is more prominent in the infrared partly because of its greater size and partly because, being darker, it absorbs more sunlight and is warmer. The full theory of this will be developed in the section devoted to asteroids. The temperatures of the Galilean satellites range from 134 to 160 °K, according to the optical albedo. At 3 and 5 μm I has an unusually high albedo,[16] and this is an extension of the optical where the satellite appears distinctly red.

It is possible to measure the satellites as they pass into eclipse. From the temperature changes we can hope to determine γ and hence decide whether these bodies are dust-clad or simply bare rock. In view of the fact that Phobos has a dust coating we can reasonably expect dust to lie on the Galilean satellites, which are much bigger and will not lose their dust so readily. The measurement involved is very difficult, especially since the satellites cannot be seen optically during eclipse and must be set on to and tracked by dead reckoning. Recent results show that all four have values of γ much higher even than that of the moon, in the range 3,000 to 4,000.[17] Further, the best fits to the observations require a two-layer model in which there is a few millimetres of powdery material of very high γ overlying a solid layer of low γ. This is an interesting result because we can estimate the effects of meteorite impacts on the surfaces of the Galilean satellites: they should fragment the top few centimetres. There should be no solid material 1 mm down. How can we explain this discrepancy?

There is one material alone which can heal its fractures: ice. The eclipse results are therefore used as evidence that the Galilean satellites are ice-covered. Indeed these were the first such evidence, presented early in 1973. By the end of that year 1 to 4 μm spectra had been taken of the four satellites and these spectra gave much more direct evidence for the presence of ice.[18] In the 1 to 4 μm region ice has some absorption bands, the most prominent being at 3·2 μm. We shall meet this band again later in the book. The ice bands were found strongly in absorption in the spectra of Europa and Ganymede. IV has weak ice bands but I has almost unit emissivity throughout the 1 to 4 μm region. Some people now talk of deposits of sulphur on Io giving rise to the high red albedo. The infrared observations also show that there are no significant ammonia or methane atmospheres to the Galilean satellites. Had we eyes sensitive in the 3·5 μm band we should see these four bright objects orbiting and being occasionally eclipsed by an invisible companion. Such observations would have been much prized by proponents of the epicyclic Ptolemaic universe!

SATURN

In most respects Saturn resembles Jupiter. It too has molecular absorption bands in the PbS region, and an atmospheric window at 5 μm. Being farther from the sun Saturn is cooler than Jupiter: 95 °K is a representative figure for the 10 and 20 μm brightness temperature. At the time of writing no maps of the planet have been attempted at any infrared wavelength. It is likely, however, that at 5 μm the results will resemble those of Jupiter. Indeed 5 μm may be a better wavelength than the visible at which to observe the belt structure of Saturn and determine the planet's rotation period.

The 7·5 to 13·5 μm spectrum was examined by Gillett and Forrest who found the ammonia bands weak and, instead, evidence of phosphine (PH_3).[19] The 7·9 μm methane band is bright, and this suggests that there is a temperature inversion as in Jupiter's atmosphere. Ethane (C_2H_6) may be responsible for an emission feature at 12 μm.

The case for an internal energy source is less convincing for Saturn than Jupiter, and it depends on observations beyond 25 μm where calibrations are not reliable.[20] In view of the other similarities, however, it is quite likely that Saturn is radiating more energy than it absorbs from the sun, perhaps in the same proportion as Jupiter.

The Rings

In much the way that we can determine the molecular composition of Jupiter and its satellites from the absorption bands in the reflected sunlight, we can make the same kind of analysis for the particles in the rings of Saturn. Though we may be unable to determine all the constituents, some at least should be identifiable. The spectrum from 1·2 to 2·5 μm has been studied by two teams who both found that ice bands dominate the spectrum.[21] Thus one of the constituents of the rings is common ice.

The thermal radiation of Saturn's rings was first detected by Allen and Murdock.[22] Because the ring particles have high optical albedoes they have a very low temperature – not much over 80 °K. Thus they are difficult to detect in the 10 μm region but relatively easy at 20 μm.

The Satellites

Titan, the principal satellite of Saturn, is the only one in the solar system known to possess an atmosphere. As we have seen, the presence of an atmosphere can play havoc with simple attempts at predicting the infrared behaviour of a planetary body. So it is with Titan. At 12 μm Titan is twice as bright as it would be had it no atmosphere.[22] The atmosphere acts as a greenhouse, letting solar radiation in but inhibiting reradiation at infrared wavelengths.

Titan's surface is therefore raised above the simple equilibrium value by about 10 °K. At 12 μm there is an atmospheric window allowing radiation from the surface to escape. This window is more opaque at 10 μm where Titan is a difficult object to observe. At 7·9 μm the methane band makes Titan bright again. Titan, like Jupiter and Saturn, may have a temperature inversion.[23]

The other of Saturn's satellites which has attracted much attention is Iapetus. Between its eastern and western elongations Iapetus changes in visual brightness by almost a factor of ten. The 20 μm magnitude has a considerably smaller amplitude and is of the opposite phase, being brightest when the visual magnitude is faintest. This sort of behaviour we have come to recognise as the effect of albedo. Iapetus is a harlequin satellite: one side is almost black, the other white. The dark hemisphere reflects only 3 or 4 per cent of the incident sunlight and as such is the blackest surface known in the solar system.

URANUS, NEPTUNE AND PLUTO

Because of their great distances from the sun and earth, the three outermost planets are rather small and very cold. The combination of the two factors makes their thermal reradiation energy very weak, and to date Pluto has not been detected at any infrared wavelength.

As long ago as 1966 Low managed to make a rather poor detection of Uranus at 20 μm, from which he deduced the surface temperature to be 55 °K.[24] This result has since been confirmed by 24 μm measure-

TABLE 3

Infrared Magnitudes of the Planets

	L	N
Sun	−28	−28
Moon	up to −15	−11 to −23
Mercury	up to −9	0 to −12
Venus	−4	−12
Mars	−4	−10
Jupiter	invisible	−7
Saturn	invisible	−3
Uranus	invisible to +2	(+15)
Neptune	invisible	(+16)
Pluto	(+12)	(+11)
Ceres	+7	−4

ments from Mauna Kea.[25] The Mauna Kea team also measured Neptune and deduced a temperature of 57° K. Both of these planets have hydrogen in their atmospheres; the measurements refer to the upper atmospheric layers because molecular hydrogen has a strong absorption feature around 20 μm.

At 3·5 μm these planets should be seen by reflected sunlight and, if methane is a major constituent of their atmospheres, should be very faint. Ney and Maas indeed could not detect Uranus at L on most of the occasions they attempted to measure it. Sometimes, however, they found a signal.[26] The presence of a strong signal related to the rotation period of the planet, and on one occasion they even recorded the signal fading away, presumably when whatever was causing it was carried out of sight by the planet's rotation. These observations were made in 1969, and since that time no enhancement of Uranus' 3·5 μm signal has been recorded. No explanation of the phenomenon has been proposed.

Table 3 lists the 3·5 and 10 μm magnitudes of the planets. Figures in parentheses are estimates where no data exist.

THE ASTEROIDS

Asteroids are quite bright long-wavelength sources. At 10 μm, for instance, several exceed magnitude -3 which is about as prominent as is a second-magnitude star in the visible. If we had 10 μm sensitive eyes, therefore, we would record many more 'wanderers' in our sky. Not only are the infrared data easy to obtain: they are also very valuable, for they enable us to estimate the diameters of the asteroids.

A number of popular books contain lists of the diameters of asteroids, and it is worth digressing for a few paragraphs to describe how these sizes have been deduced. The first four asteroids, (1) Ceres, (2) Pallas, (3) Juno and (4) Vesta were measured around the turn of the century by means of micrometry. The discs of these asteroids are a few tenths of a second of arc in diameter, just large enough to resolve with very large telescopes. The micrometric technique involved setting the wire of a micrometer first on one side of the disc, then on the other, and measuring the displacement between the two positions. Considering the effects of atmospheric turbulence on an image of that size and the fact that the thickness of the micrometer wires themselves was not much less than the size of the asteroid disc, we could hardly expect the accuracy of such measurements to be high. Yet the figures determined in this way, principally by Barnard, are in use to this day.

The diameter leads to the albedo of the asteroid – the proportion of sunlight which is reflected back towards us. We must guess how much is reflected in other directions. A simple model has the brightness of the asteroid dependent on the albedo, A, the square of its diameter and the distance from the sun and earth. The values of A derived from the micrometric diameters ranged from 0·03 to 0·27. Although spanning an order of magnitude, these four albedoes were averaged together, and the resulting mean was taken to be an appropriate value of A for other asteroids. Hence from any observed brightness and the assumed albedo the diameter of any other asteroid could be – well – guessed. For it was nothing more than a guess, and a guess based on data which were themselves of questionable reliability.

Today there are three independent ways of determining the diameters and albedoes of asteroids, and all three seem to be agreeing quite closely. The most direct is by occultations. If an asteroid occults a star, measurements of the duration of the occultation from several points on the earth's surface lead to a very accurate determination of the asteroid's diameter. Such occultations are, however, very rare. Occultations of the asteroid by the moon also give a diameter if high-speed photometry is used, for it takes up to half a second for the asteroid to disappear. This technique can be used only for the brightest asteroids, and occultations of these by the moon are also very rare.

The second method involves measurement of the polarisation of reflected sunlight. The details of this method will not be expounded here except to remark that it leads to a fairly reliable determination of A and hence of the diameter.

Of the three 'modern' methods of determining the diameters of asteroids, infrared photometry was employed first. The method is basically simple. If a fraction A of incident sunlight is reflected, the remainder, $1-A$, must be absorbed. Knowing how much energy is absorbed we can calculate the temperature by equating the absorbed energy to σT^4. And from the temperature we can calculate the infrared flux to within a factor of d^2, d being the diameter we seek. In effect we have the product $d^2(1-A)$ from the infrared brightness and d^2A from the visible brightness. These two numbers can be manipulated to give both d and A.

The infrared method was first used by Allen in 1970 on the asteroid Vesta.[27] He found a diameter 50 per cent larger than the quoted micrometric value, and at first this result was not accepted by many astronomers used to more orthodox methods; it was almost two years before confirmation came. Many asteroids have now been measured, principally by Matson for his PhD thesis at CalTech and by David Morrison of the University of Hawaii.[28] Some of the results are listed in Table 4. It now seems that all the original micrometric measures were low by a significant amount. To date three asteroids have been found which are extremely dark and therefore much larger than once thought: these are 19 Fortuna, 324 Bamberga and 747 Winchester. It is likely that the infrared method will result in the discovery of more of these low-albedo objects whose existence was previously unanticipated. The infrared method has also been used on some of the satellites of the outer planets where it gives values in agreement with optical determinations.

In this brief treatment of the infrared method, many assumptions will be glossed over, but two in particular should be mentioned for if badly wrong they could introduce errors of 10 or even 20 per cent into the deduced diameters. Since, fortuitously, they act in opposite senses, the possible errors tend to cancel.

The first implicit assumption is that the surface of the asteroid is Lambertian. Since we know that the moon and Mercury are not Lambert spheres, it seems unlikely that asteroids are. If the moon is a reasonable yardstick, asteroids may emit 10 per cent more than expected because of surface roughness. In all probability even 10 per

TABLE 4

The Diameters of Some Asteroids
Determined by Infrared Photometry

Asteroid		Infrared diameter (km)	Micrometric diameter (km)
1	Ceres	1040	740
2	Pallas	570	480
3	Juno	250	200
4	Vesta	540	380
5	Astraea	130	
6	Hebe	220	
7	Iris	210	
8	Flora	170	
9	Metis	210	
10	Hygiea	420	
15	Eunomia	270	
16	Psyche	270	
19	Fortuna	230	
433	Eros	25	
511	Davida	310	

cent is too low a figure for its effect on the smaller asteroids which may bear no resemblance whatsoever to spheres.

The second assumption is that rotation is unimportant. If the asteroid rotates, as indeed is almost certain, not all the energy absorbed on the sunward side is reradiated back towards the sun and earth. Some must be radiated from the night hemisphere, but just how much depends on the rotation period, which is usually a few hours, and on γ. If γ is large, as for the moon and the Galilean satellites, only about 10 per cent of the incident sunlight is radiated from the dark hemisphere, but if the asteroid is naked rock with no covering of dust, as much as 30 per cent may escape in that way.

It is unlikely that we will learn sufficient about the thermal inertia and surface roughness of asteroids by earth-based study to improve on the infrared method. And if we must rely on observations by space probes the purpose of the exercise has been defeated. Rather, as accurate diameters are found by one of the other two techniques, the infrared photometry might be used to determine the roughness and thermal inertia of asteroids. Also, and perhaps more instructive, application of the infrared method to asteroids with large periodic optical variations will tell us whether these variations are caused by changes in projected shape or in albedo as the asteroid rotates.

COMETS

Only a small number of comets has been observed in the infrared and in this section three will be described – the bright comets of 1965, 1969 and 1973–4.

Comet Ikeya-Seki 1965f

Southern hemisphere observers will recall this spectacular sun-grazing comet which was so prominent an object in the dawn sky in October 1965. Infrared astronomers remember it because it was the first comet to be observed in their wavelength domain. Only one group published observations of 1965f, Eric Becklin and Jim Westphal of CalTech; they used a 24-inch reflector on Mount Wilson.[29]

The approximate energy distribution of 1965f is shown in Fig 13. This is in fact a typical curve for solar-system bodies. On the left the falling portion of the curve is reflected sunlight. The portion at right is the thermal reradiation, roughly resembling a black body. The presence of this reradiation component tells us that there is solid material present, for only a solid can absorb sunlight and re-emit the energy in this way. But a comet is not a simple solid object like a planet or satellite, and this is proved by the fact that the thermal radiation originates not just in the nucleus but comes from the entire head. Indeed radiation was traced by Becklin and Westphal for a considerable distance along the tail.

Becklin and Westphal correctly concluded that the thermal radiation arose not from a single solid body but from a distribution of small solid particles. These particles must be tiny meteoric morsels, in a word *dust*. Dust is certainly known to be present in the tails of comets, for we can see there the haze of sunlight reflected off all the tiny particles. So it is not surprising that infrared observations recorded the presence of it in the head.

Since almost the entire remainder of this book will deal with the occurrence of dust and observation at infrared wavelengths of its thermal radiation, it is worth detouring for a sentence or two to state in a little more detail what is involved. The dust we observe in the infrared is probably very small. The particles may be as little as a micron across – one thousandth of a millimetre. Being so small they are isothermal: they have the same temperature throughout and therefore give the same signal from whichever side they are viewed. Like all solids they melt, or in the vacuum around stars, evaporate when the temperature gets uncomfortably high. Dust cannot exist above about 1,500 °K therefore. What the chemical composition of the dust is – well, that won't be fully answered even by the end of the book.

From Wien's law, Becklin and Westphal could deduce the temperature of the dust. At the closest to the sun they were able to observe 1965f, the dust had reached almost 1,150 °K. Had they been observing a large object at this temperature, large enough to fill the beam, they would have recorded a signal more than one thousand times stronger. Thus they could deduce the *optical thickness* of the dust to be about 10^{-3}. In other words only one thousandth of the projected area of the comet was actually in the form of dust grains, and this means that individual grains were a considerable distance apart.

The equilibrium temperature of black-body dust grains at the appropriate distance from the sun was 750 °K when the empirical temperature was 1,150 °K. To account for the discrepancy, Becklin and Westphal postulated that the grains were more efficient at absorbing sunlight than at emitting infrared radiation. If a grain cannot emit easily, because its emissivity is low, it must settle to a higher temperature where, by Stefan's law, it will compensate. The variation of emissivity from the visible to the infrared required to

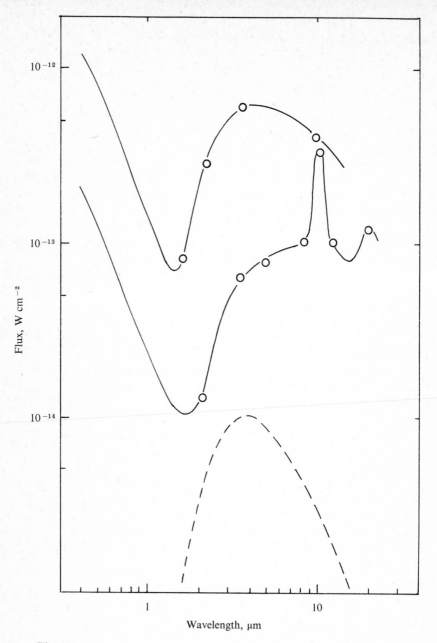

Fig. 13 Energy distributions for comets Ikeya-Seki and Bennett. This is a logarithmic plot of the flux against wavelength, and a 1000 °K black body is shown for comparison. The upper curve is Ikeya-Seki, the brighter of the two; the lower is Bennett.

explain the discrepant temperatures was closely matched by that of iron. They therefore suggested that iron might be a major constituent of comets. Since iron is a common mineral in meteorites, and since meteorites are associated with comets, this seemed an entirely reasonable identification.

Comet Bennett 1969i

Maas, Ney and Woolf used the 30-inch Minnesota telescope to observe comet Bennett in 1969.[30] Whereas Becklin and Westphal had used a broad 10 μm filter on 1965f, the Minnesota photometry employed narrow-band filters in the 8–14 μm window. Their results are also shown in Fig 13. As for 1965f, 1969i had a colour temperature in excess of its equilibrium value. Maas *et al* commented that this could be explained as well by carbon particles – graphite – as by iron. The optical thickness of 1969i was slightly greater than that of 1965f: it contained more dust.

It can be seen in Fig 13 that the data around 10 μm do not fit a simple black-body curve. The strange bump at 10 μm can be explained only by an emissivity effect, and indeed similar features had already been found in the energy distributions of some stars, as Chapter 5 will reveal. A narrow emissivity feature such as this is a very distinct clue to the composition of the material emitting. Studied at sufficient spectral resolution it is as clear as a fingerprint. The feature which appeared at 10 μm in comet Bennett was caused by grains of silicate dust.

Silicate dust is not a surprising thing to find. Many meteorites, the stony variety, are composed almost entirely of silicates. Most of the earth's mantle is too, from the sandstones of Grand Canyon to the soil in your back garden. Probably the entire moon is composed of the silicates of various metals. It now seems only natural that comets should contain silicates, but in 1969, when 10 μm 'bumps' had just begun to come to light, this discovery by the Minnesota team was a significant result which linked the chemistry of our own solar system with that of other stars and their circumstellar environs.

Comet Kohoutek 1973f

The Christmas comet of 1973 was something of a flop for the observers who relied on small pairs of binoculars for its study. In the infrared it was quite prominent, and its magnitude at 20 μm, in excess of −6, at least sounded impressive. Like comet Bennett, 1973f had a silicate bump, albeit smaller, and, as always, its colour temperature was too high.

Observations were made by Ney who was able to measure not only the head and tail, but a beard (or anti-tail) as well.[31] The tail had the same colour temperature as the head, and the same silicate bump. The beard was quite different: there was no hint of a silicate feature and the colour temperature was almost exactly the equilibrium value.

Clearly the beard of comet Kohoutek contained different particles from the head and tail, but different in what way? There are two possibilities: the difference could be in composition or in size. Of these Ney favoured the latter. If the particles of the beard are much larger than those of the head they will settle to a temperature closer to the equilibrium value. Further, the silicate emissivity feature is destroyed if the grains are much larger than the wavelength, 10 μm. Thus both observations are accounted for simultaneously.

Rieke and Lee also measured comet Kohoutek and concluded that it contained no more than one-sixteenth as much dust as did comet Bennett.[32] This, they reasoned, was at least a partial explanation for why it did not become as bright in the visible as early predictions suggested it might.

No attempt has been made to detect meteors as they pass through the atmosphere. In theory the brighter naked-eye meteors should radiate sufficiently strongly to be detected, and it is not inconceivable that meteors account for some of the spikes that one finds on observing records. However, the chances of a bright meteor passing through the tiny beams usually employed are exceptionally small. The best hope of detecting meteors is by the statistical method: aim a detector at the sky during an active meteor shower and compare the number of spikes recorded in the night with the number on a non-shower night.

One very interesting observation of a meteor has been made from an infrared satellite looking down at the earth. The occasion was August 10 1972, and this observation was reported in 1974.[33] The meteor came in at a low angle and instead of ploughing down to the ground as most do, it actually bounced off the atmosphere and continued its journey having done no more than ruffle its feathers. The meteor was very bright and was seen over much of the USA and Canada. From the infrared measurements the diameter of the meteoric body was estimated at 4 metres; if this is correct it was one of the largest objects the earth might have encountered in 1972. It was, perhaps, just as well that it didn't land anywhere.

THE ZODIACAL LIGHT

The zodiacal light is produced by sunlight scattered off dust grains. Since these grains lie quite close to the sun, they must be at temperatures around 1,000 °K, and should therefore emit strongly in the infrared. Their detection is one of the most difficult jobs one could attempt, however. For a start the optical thickness of the zodiacal light is exceptionally small, so that the signal is weak. Secondly the zodiacal light is so extended that it is impossible to chop with one beam on the light and one beam off it. Finally, near the sun where signals should be larger there is the problem of excessive scattered radiation which introduces an additional component of sky noise at short wavelengths.

Attempts have been made to measure the outer corona and inner zodiacal light during total eclipses of the sun. At wavelengths up to 2·2 μm this is not too difficult because it is possible to chop against a room-temperature black body instead of against a reference sky beam. There is no indication of dust at 2·2 μm as far as has been measured.

An ingenious way of measuring the zodiacal light at longer infrared wavelengths was devised by Ney. He waited for the moon to pass close to the sun, then set his telescope so that one beam was on the dark side of the moon and the other on the sky. In this way he hoped to find the sky to be brighter than the moon. But not even this yielded a measurable signal and he could place only an upper limit on the infrared flux.

Rocket-borne infrared detectors have succeeded in measuring the very weak signal from the zodiacal light at 13 and 20 μm at a fairly large elongation from the sun.[34] The colour temperature, 280 °K, was close to the expected value for the observed position, but the optical thickness was around 3×10^{-8}, a value much smaller than that of the most tenuous of comets.

5

COOL STARS

The CalTech 2-micron survey, the IRC, contains mostly cool stars of spectral type later than KO. This of course comes as no surprise, for while the energy distributions of hot stars, at temperatures around 10,000 °K, are falling throughout the infrared, those of cool stars at about 3,000 °K rise to a peak just short of 2 μm. As an example we may consider the constellation of Orion. In the visible Rigel and Betelgeuse are approximately equally bright. At 2 μm the magnitude of Rigel is little different at $+0\cdot2$: this is because both wavelengths lie on the Rayleigh-Jeans tail of Rigel's black-body curve. Betelgeuse, however, has risen to become the brightest star in the sky, the sun apart, at magnitude $-4\cdot5$. Thus almost a factor of 100 separates the two.

Because cool stars are so bright at infrared wavelengths, they were the first to be studied extensively. Still the greatest body of data is that provided by the IRC, and only a small proportion of the many cool stars contained in it have been studied in much detail at other wavelengths. At about the same time that the first IRC results were emerging from the mountain of data, interesting observations were being made at 10 μm, principally from Minnesota.

COOL STARS IN THE INFRARED CATALOGUE

Many of the cool, late-type stars in the IRC have more or less black-body energy distributions. Just what percentage we do not know. A sizeable proportion certainly do not follow the Planck curve, and for some of the remainder the colour temperature is disproportionately low. This, in fact, is the basic finding of the 2-μm survey.

The first objects published, known as NML Tauri (now IK Tau, for it is a variable star) and NML Cygni, have particularly low colour temperatures. Fig 14 compares the energy distribution of these two with a normal cool star (a Tau) and a hot star (a Lyr). NML Cyg is particularly bright, being magnitude -8 at 20 μm, and was the

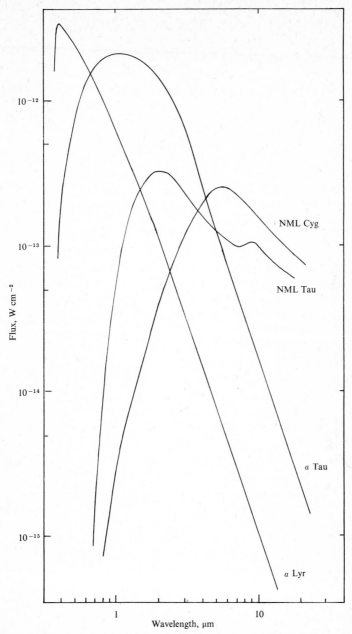

Fig. 14 Energy distributions for the infrared sources NML Cygni and NML Tauri. Two normal stars are shown for comparison: α Lyrae (Vega) is very hot and α Tau (Aldeberan) is a cool star.

77

subject of a number of speculative papers by various authors in the mid 1960s. All agree that no normal star should have an energy distribution like that of NML Cygni: some extra mechanism is contributing to make the star brighter in the infrared. It has what we call an *infrared excess*.

The term infrared excess has become so over-used as to be rendered nearly meaningless. Almost every class of object in the sky contains members which can be said to have an infrared excess: the cliché is exercised whenever the observations lie above an extrapolation of the optical energy distribution into the infrared. But what this really means is not that there is something special about the objects concerned; rather it is a poor excuse for our inability to predict their behaviour at infrared wavelengths. In the early days of infrared photometry this was all well and good, but now that our understanding has deepened considerably there is really no point in referring to infrared excesses in objects one fully expects to exhibit a second emission mechanism in the infrared. Nonetheless, the term has a certain ring of authenticity, even adventure, to it. You can expect more people to read your paper if the word 'excess' appears in the title than if you put something about the usual infrared emission mechanism. Besides, 'infrared excess' is probably the shortest way of getting over the message. Thus there is no likelihood of its being dropped from the infrared astronomer's vocabulary in the near future. And so long as everybody is clear in their minds that in the vast majority of cases it just refers to the usual humdrum emission mechanism, then probably no harm done.

So NML Cygni has an infrared excess, and the next question is what causes it? We do not have to look far to find the answer: the comets have already told us. The excess is caused by thermal radiation from circumstellar dust grains.

The mechanism is the same as for solar-system bodies. Each grain soaks up the starlight that hits it, settles to an appropriate equilibrium temperature below about 1,500 °K, and radiates the energy thermally. When we examine just how much dust radiation comes from a star like NML Cyg, we realise that we are dealing not with one or two comets or meteor swarms around the star but with massive clouds of circumstellar dust and gas. So thick can these dust clouds be that the star is virtually extinguished in the optical just as the sun can be by very thick smoke from a chimney. Not only does the dust raise the infrared flux by factors of several thousand or more, it also dims the optical and gives the impression of an enormous infrared excess.

Stars with large infrared excesses are sometimes called *cocoon stars* and described as having *dust shells*. This may be appropriate to NML Cyg itself, but it conveys the impression of a star entirely surrounded by a spherical and uniform envelope of dust and gas. Such a picture is certainly misleading for many stars. In this book the term *dust cloud* is employed deliberately to make the concept flocculent and to convey the impression of localised patches of dust with clearer regions between.

What sort of star is NML Cyg underneath its dust clouds? A number of theories were evolved, some authors thinking it to be a very young star recently formed and not yet settled into a stable life on the main sequence, others thinking it to be an old late-type object. That the latter theory is correct was demonstrated by Johnson who assigned to it the spectral type M6 Ia.[1] Class Ia are the most luminous supergiants known, and M6 is just about as cool as they are found. So NML Cygni is an enormously big, bright but very cool star surrounded by a dust cloud.

There is certainly dust around NML Tauri, but the underlying star is quite different from NML Cyg. It varies with a period of about 500 days, and now appears in the *General Catalogue of Variable Stars* as the star IK Tau, Mira type variable. It is best known by its original designation. The spectral type is also M6, but it is only a giant.

Steadily the interesting objects filtered out from CalTech. In 1966 the CIT sources were released, and photometry to 3·5 μm was published in 1967.[2] Curiously these have been ignored at longer wavelengths until very recently. Most have dust envelopes, but CIT 11 appears red at K because it experiences a large interstellar extinction (and hence reddening). It lies, in fact, behind the Great Rift in Cygnus. Five of the CIT sources were already known Mira type variable stars, the same type as NML Tau. These are: CIT 2 = RW And, CIT 7 = WX Ser, CIT 8 = RU Her, CIT 9 = MW Her and CIT 12 = DG Cyg. CIT 1 and 10 are invisible at optical wavelengths.

VY CMa is another interesting star released privately by Becklin and Neugebauer as early as 1968 and first described fully late in 1969.[3] In some circles VY CMa is known by the nickname 'Vicky'. It too is a star surrounded by a dust envelope, but unlike the others so far described it is bright at optical wavelengths – about magnitude 8. It lies in a small bright nebula and is listed in double star catalogues as multiple. Several companions have been measured by various double-star observers, and each has varied greatly in visibility. The motions are unusual: all the companions are moving radially out-

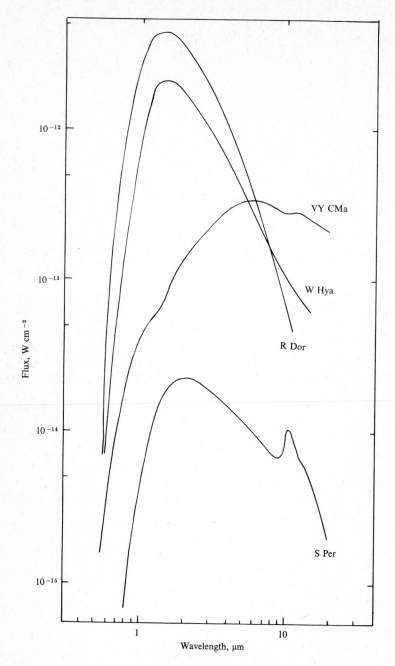

Fig. 15   Energy distributions for four more infrared sources.

wards from VY CMa. In fact these *comites* are not stars at all but small dense knots in the nebula.

The distance and luminosity of VY CMa are not known, and there is some debate about its nature. Is it, like the NML and CIT stars, a late-type object? Or is it young? At the moment we do not know, but most authors favour the former.

Another bright late-type star with a pronounced dust shell is S Per, an M4 Ia supergiant,[4] and two similar objects in the southern skies are R Dor and W Hya.[5] Energy distributions for these three and for 'Vicky' are given in Fig 15, on the same scale as Fig 14.

The 2 μm survey, then, showed us that many cool stars have circumstellar dust clouds of sufficient thickness that most of the total luminosity of the system is emitted by the infrared component. But the photometry we have presented so far is only broad-band data. We can learn considerably more when we split the 8–14 μm atmospheric window into narrower portions and examine even quite normal cool stars.

SILICATE BUMPS

Woolf and Ney (1969) is a classic paper in infrared astronomy.[6] In it is described an infrared excess which occurs in certain cool stars between 8 and 12 μm. This excess has a very characteristic shape, shown in Fig 16, and is sharply defined in wavelength. It occurs strongly in some stars, including a Ori and o Cet (Mira), but is quite absent from other cool stars (a Tau, a Boo). It is narrower than a black body and must therefore be an effect of the emissivity of the circumstellar dust itself. It is, in fact, the very same 10 μm 'bump' that was found in comet Bennett: it is caused by silicates. Some late-type stars are surrounded by clouds of quite ordinary silicates.

Of course the sun has a lot of silicates around it, but if viewed from a distance it would probably not show much of a silicate bump, if any. Most meteorites, asteroids, planets and satellites, even though composed of silicates, have black-body emissivities. This is because of their size. If a grain is to display its natural emissivity it must be small – smaller than the wavelength of the radiation involved. The grains around a Ori are no more than a few microns across, microscopic particles 10,000 of which could be laid between the divisions of an inch ruler. Being so small they must also be very numerous: a Ori probably has about 10^{39} of these grains in its circumstellar clouds.

Many sceptics refused to accept the presence of grains as compli-

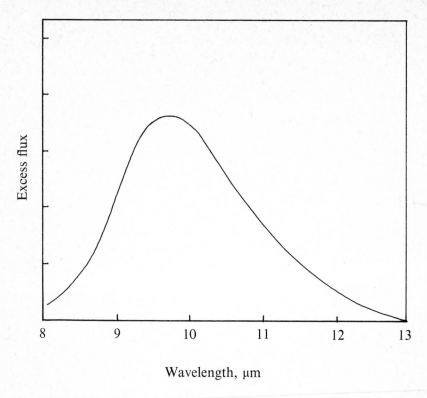

Wavelength, μm

Fig. 16 The 'silicate bump'. This is the excess which remains when the stellar continuum has been subtracted. A second bump occurs in the 20 μm window, but no accurate shape for this has been determined.

cated as silicates around late-type stars; it was well known that graphite and ice were the most likely constituents of the interstellar dust. These sceptics argued privately (but never in print) that Woolf and Ney must have been confused by the atmospheric transmission which falls off at 8 and 12 μm in much the same way that the silicate bump does. But the evidence strengthened, and when finally such a bump turned up in comet Bennett too, most of the sceptics were finally silenced. Another piece of evidence was brought into play: silicates, like some camels, have two bumps and the second very conveniently falls in the 20 μm window. Low and Krishna Swamy showed that α Ori does indeed have a 20 μm silicate bump.[7]

What sort of stars have silicate bumps? This was the question the Minnesota team addressed themselves to, and the first answer they came up with was: SRc variables do.[8] SRc (formerly Ic) variables are late-type stars of small amplitude and rather erratic light curve;

they are all supergiants. α Ori is a typical example; VY CMa is also classified SRc. The only SRc variables lacking silicates are WY Gem, KN Cas and ψ^1 Aur, and of these the first two have hot companions which very possibly destroy the dust.

This was only a start: it soon became apparent that SRc stars were by no means the only ones with silicates. The picture that now emerges is basically that the larger and the cooler the star, the more prominent the silicate bump.[9] Supergiants are therefore more likely to have a silicate bump than giants, and while stars of spectral type M0 rarely show silicate emission, later than about M4 the bumps are common.

Most of these late-type giants and supergiants display only the silicate excess; there are no black-body excesses of the sort found in some IRC stars. Out to 5 μm the fluxes fit exactly those expected of the star, including an absorption in the 5 μm band due to carbon monoxide. But theory predicts that the narrow emissivity feature is destroyed not only by the particles being too big, but also if they are too numerous. In either case the silicate feature is steadily distorted into a black body as the size or number of the particles is increased. It is of interest, then, to examine the late-type stars with black-body dust emission to see if they have vestigal silicate bumps as well. The results are varied. For example NML Cyg has not the slightest ripple at 10 μm; VY CMa shows a possible silicate bump;[10] S Per shows a weak hint of silicates.

If the silicate bump is measured at high spectral resolution there exists the possibility of identifying the exact type of silicate responsible, for each has a slightly different spectral shape. This has been often attempted but rarely published, for·it involves stretching the data too far. Besides, there is no guarantee that only one type of silicate is present. Gammon, Gaustad and Treffers examined the excesses in VX Sgr and o Cet and concluded that the types of silicates involved are basic rather than acidic, but further than this we cannot reasonably go.[11]

Although in general a star must be a supergiant of late spectral type in order to have a strong silicate excess, a few exceptions occur. Humphreys, Strecker and Ney found several very luminous G supergaints with silicate excesses,[12] and Humphreys and Ney reported silicates in HD 101584, an F2 supergiant. This star, however, they consider to be a binary in which the companion is an M supergiant.[13] They invoke the same explanation for 89 Her, an F0 supergiant with a sizeable infrared excess.[14]

Another exception is 17 Leporis, a binary star with companions an A0 dwarf and an M1 giant. This system has a pronounced silicate bump, yet there are no examples of either A0 dwarfs or M1 giants which individually have silicate emission.[15] The two stars pass very close to one another on occasions, and when this happens material may be sucked off the M1 giant and blown away by the A star. Here we find a clue to the origin of the silicate grains: they can be formed from the material of which late-type stars are made when it is driven to great distances.

Most late-type supergiants are in fact losing mass. These stars have become so obese with their change of diet in later life (burning helium instead of hydrogen) that they are forced to lose weight. This they do by allowing their outer layers to smoke gently into space. There is direct spectroscopic evidence for this mass loss, and estimates can be made of its magnitude. A good correlation exists between the rate at which material escapes from late-type stars and the intensity of their silicate bumps.[9]

To substantiate further the claim that silicates are formed from the material ejected by cool stars, Gilman calculated what solids would condense from gas of stellar composition as it moved away from a star and cooled.[16] This is critically dependent on the ratio of oxygen to carbon in the gas. These elements first combine to form CO; the subsequent development depends on which of the two is left over when all the other has been used up in this way. If carbon predominates, graphite will be the principal condensate, or under certain circumstances silicon carbide. If oxygen wins the contest, the grains which should form are the silicates of calcium, magnesium, aluminium and iron; combinations of these are responsible for the 10 and 20 μm bumps.

Oxygen-rich late-type stars frequently turn out to be OH sources too. OH is one of those molecules that shouldn't exist. It is trying hard to be water but failing because insufficient hydrogen atoms have crossed paths with each oxygen atom. We observe this half-naked oxygen at radio wavelengths around 18 cm where there are four emission lines produced by various subtle transitions. The conditions under which OH emission is excited are not yet quite certain. Clearly OH molecules must be present, and this requires an oxygen-rich gas. Secondly some source of energy must exist which constantly excites the molecule to its higher energy state: from this precarious perch the molecule topples, emitting its distress call at suitable radio wavelengths. The pumping mechanism which keeps forcing the reluctant molecules round this perpetual cycle is almost certainly radiation at one of several possible infrared wavelengths. So to

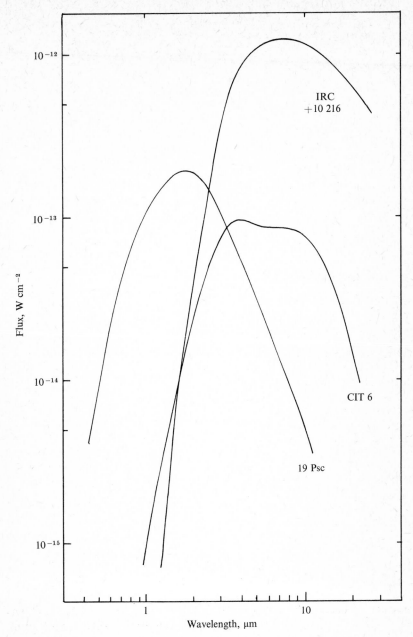

Fig. 17   Energy distributions of three carbon stars. The infrared continua of these stars are generally smooth and do not show silicate bumps. 19 Piscium is the variable star TX Psc.

produce OH emission we need infrared radiation, and what better source than late-type stars?

Late-type stars; oxygen rich; circumstellar gas: it all adds up to a correlation between OH sources and cool stars with silicate excesses and/or thick dust clouds. This correlation is indeed found.[17] NML Cyg is one of the strongest OH sources known, and a similar object is IRC+50 137, a source too faint to be recorded on photographs at optical wavelengths. H_2O emission also frequently occurs in cool stars with silicate bumps. Like the OH molecule, water has a number of emission lines at radio wavelengths.

CARBON STARS

Gilman's calculations suggested that cool stars with more carbon than oxygen might produce circumstellar clouds of graphite or silicon carbide grains. Such stars can be recognised spectroscopically by their molecular absorption lines of CO and CN. They are allotted the spectral types R and N and are known collectively as carbon stars.

Graphite has no quirks of emissivity; ε falls slowly to longer wavelengths and the presence of graphite grains around a star should be manifested by a roughly black-body excess. SiC has an emission feature in the 8–14 μm window differing in shape from that of the silicates. The number of known carbon stars is not very great, and most of them have been observed at infrared wavelengths. Of these a sizeable proportion do indeed have black-body excesses. The first such excess to be described was that of R Lep;[6] the colour temperature of the excess, determined from Wien's law, is about 140 °K in this star. Further observations suggest that this may be a SiC bump rather than a cool black body. Energy distributions of several carbon stars, some with small 10 μm bumps attributable to SiC, are shown in Fig 17.

"CYGNIDS AND TAURIDS"

When Strecker, Ney and Murdock published a paper describing infrared observations of "Cyngids and Taurids", the first thought of this author and others was that they had at last detected meteors from two of the major optical showers. What they were in fact doing was dividing the cool stars with infrared excesses into two groups with NML Cyg and NML Tau as type objects.[18] Examining a selection of 224 IRC sources they found that over 80 per cent had no appreciable 10 μm excess. Those with excesses divided roughly as two resembling NML Tau to every one resembling NML Cyg. The characteristic energy distributions of the two types are shown in Fig 18.

NML Cyg and the stars which resemble it have roughly black-body excesses. Most of them are variable, the variations being

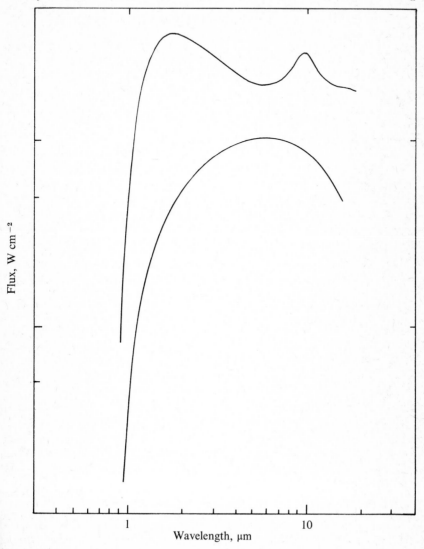

Fig. 18   The two types of infrared continua in late-type stars. The upper curve is typical of an NML Tauri star; the lower of an NML Cygni star.

greater at the shorter wavelengths than the longer. Thus it is the temperature of the dust which varies. The stars in this class are of a variety of types. Some are carbon stars; some are oxygen-rich stars surrounded by optically thick silicate dust that has adopted its blackbody form. One or two are stars which we will meet in the next two chapters.

The stars like NML Tau form a more uniform sample. They are all oxygen-rich giants or supergiants with typical silicate bumps. When variable the variations are independent of wavelength throughout the infrared. The continuum underlying the silicate bump is not that of the star, however, for it falls too slowly with wavelength. Whereas the Rayleigh–Jeans tail falls as λ^{-3}, the continuum from 2 to 8 μm in NML Tau-type stars drops as λ^{-1}. There are a number of plausible explanations for this; the one favoured by the Minnesota team is free-free radiation.

Free-free is a term used primarily by astronomers who do not enjoy getting their tongues round the German name *Bremsstrahlung*. It is a form of radiation produced by a partially ionised gas. When electrons and positive ions coexist in a gas, their paths frequently cross. The electrons, being lighter, orbit around the ions in much the same way that comets journey through the galaxy making a single pass of each star they encounter. Every time this occurs the electron loses a little energy and emits a photon of infrared or radio radiation. We call this free-free radiation because the electron starts free and ends free. This process cannot go on indefinitely, however, because the electron has only a limited supply of energy to spare. sooner or later it will radiate one too many photons and collapse exhausted into a closed orbit about the ion, neutralising the ionic charge. And there it will sit until something comes along to ionise it. So there must be a source of energy capable of ionising atoms, and the energy thus input cannot be less than the free-free radiation emitted by the gas.

The spectrum of free-free radiation is shown in Fig 19. It can be divided into three portions. On the left, at the shortest wavelengths, is a rising portion rather resembling the Planck function. Like the Planck function the position of this depends on the temperature. Two cases are shown, one for a hot gas at 10,000 °K and the other for a cooler gas at 3,000 °K. The temperature is determined principally by the mechanism which ionises the atoms. If electrons are given a hefty kick, as by the ultraviolet radiation of a hot star, the temperature of the gas will be high. If the electrons are just tapped lightly out of the atoms the temperature will be low. We will find

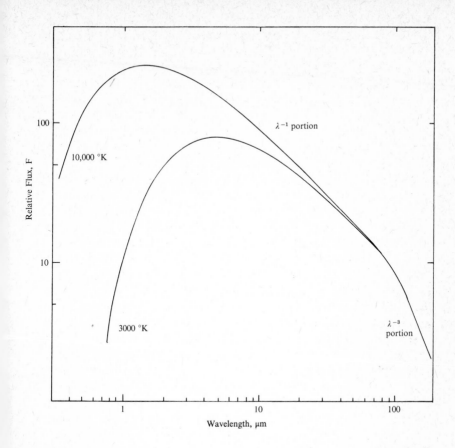

Fig. 19 The energy distribution of free-free radiation (Bremsstrahlung) from an ionised gas at 3000 °K and 10,000 °K.

examples of free-free emission at both temperatures in this and the next chapter.

The middle section of the free-free curve has the λ^{-1} dependence on wavelength. If $F\nu$ is the ordinate, as is usual in radio astronomy, this portion is a constant independent of wavelength. The vertical scale, the flux of the free-free emission, is determined here by the volume of the gas cloud and the number density of electrons and ions. The final portion of the free-free curve, which usually occurs in the radio region, falls as λ^{-3}. Here the gas has become optically thick so that free-free photons cannot escape in a simple way but are reabsorbed. The energy distribution then reverts to its natural black-body state, in this case part of the Rayleigh-Jeans tail. The position at which the turnover from λ^{-1} to λ^{-3} occurs is determined by the number density of electrons. The higher the density the shorter the wavelength of the turnover.

Returning now to the NML Tau stars we see that the portion of the energy distribution in the PbS region is nicely fitted by free-free radiation at the lower of the two temperatures we have considered. Of course, the simple fact that it appears to match is no proof that free-free really is the mechanism. We must also show that we would expect free-free radiation to be present. In other words we must find a mechanism to ionise enough gas to the right sort of temperature. Gilman thinks this is possible.[19] In his model the gas is ionised by the convection that occurs at the star's surface. We believe that the outer layers of cool stars like NML Tau are churning over with much violence. If this is so, the activity could send shock waves out into the circumstellar gas, and these waves could probably supply the energy required by the free-free interpretation.

There is one extra piece of evidence to support Gilman's argument. In his model there are rather few positive ions, and the *Bremsstrahlung* occurs between the electrons and neutral hydrogen atoms. Adultery is no sin in the quantum jungle: even a proton happily married to an electron (to form a hydrogen atom) can attract another electron who passes close enough. Essentially this is because one electron cannot be on all fronts at once to defend his precious proton against the advances of others. Where *Bremsstrahlung* occurs between neutral hydrogen atoms and electrons there will also be a little radiation at around 1 μm. This is *free-bound* radiation produced when the electron is finally captured. In the NML Tau stars there is indeed a small additional bump in the energy distribution at 1 μm.

IRC+10 216

In the summer of 1969 a conference was held in Cambridge (England) on the subject of infrared and microwave astronomy. At this conference a rather unassuming Australian, 'Harry' Hyland, then working at CalTech, presented a paper. He began: 'I want to announce the discovery of the brightest celestial 5 μm source.' The source he referred to was IRC+10 216.

IRC+10 216 is not a star. It is a tiny elliptical nebula in Leo, about 2″ arc across and 19th magnitude in the visible.[20] Its energy distribution is included in Fig 17. That the underlying object is a carbon star was shown by Miller.[21] IRC+10 216 is the carbon star with the most extreme infrared excess known.

It so happened that in the winter of 1970–1 IRC+10 216 underwent a series of occultations by the moon. These provided a very accurate way of determining the diameter of the dust cloud, something which can rarely be achieved by other techniques. The signal was monitored at several wavelengths as the moon swept over the source. It took about one second for IRC+10 216 to be occulted, and the time required was longer at 10 μm than at 2 μm.[22]

At 2 μm we record primarily hotter dust than at 10 μm. The hotter dust must lie closer to the star, thus we would expect the object to appear smaller at the shorter wavelengths, as observed. At 10 μm IRC+10 216 is a little bigger than the optical image on the best available photographs.

RV TAURI STARS

The IRC does not go very deep, and there are many stars with infrared excesses which did not quite make it into the 2 μm survey because they were just too faint at K. In the remainder of this chapter we will examine a number of types of cool stars with unusual infrared behaviour, and while some of them are in the IRC, others equally interesting are not. The RV Tauri stars are an example of the latter, for only one or two are IRC stars, and these are the least distinguished of their tribe.

It all began when Bob Gehrz, then a Ph D student at Minnesota, in a spare moment examined AC Her. He found it easier to measure at 10 than at 3·5 μm. Stars for which this is so form a very select group, so his interest was immediately aroused. In the variable star catalogue AC Her is classified as an RV Tau star; these are yellow supergiants (spectral types G–K) with fairly regular periods but variable depth of their minima.

Gehrz proceeded to examine all the RV Tau stars accessible from Minnesota and found more than half of them to have infrared excesses.[23] The form of the excess differed from silicates, graphite and silicon carbide. The excess is more common in the RV Tau stars which are known from optical spectroscopy to be metal-rich – ie the stars with more carbon, silicon, oxygen etc.

R CORONAE BOREALIS STARS

R CrB is the brightest member of a class of variables with rather unusual light curves. For most of the time these stars vegetate quietly at maximum light. Then, dramatically, they fade by up to seven or eight magnitudes. The fading takes place quickly; the subsequent brightening is slower. Often there are two or three minima. Spectroscopically the R CrB stars are distinguished by possessing little hydrogen and much carbon. It was therefore believed that these stars spasmodically puff out material from which graphite grains condense to cause the sudden fade. As the graphite moves away from the star and is dispersed, the star brightens again.

If R CrB stars have graphite grains around them, they should have black-body excesses. In 1969 just such excesses were found for two of these stars: R CrB itself[24] and RY Sgr[25]. More recently Feast and Glass found infrared excesses in an extensive sample of R CrB stars.[26] v Sgr also has an infrared excess:[27] this star spectroscopically resembles the R CrB stars but is not variable.

At minimum light, when a cloud of graphite is formed, the dust should be closest to the star, and warmest. As the cloud is dispersed the dust must cool, and this would be manifested by a longward shift in the peak wavelength of the infrared component. At the same time the intensity of the infrared component must decrease, for as the star appears to brighten less energy is being absorbed by the dust. Forrest, Gillett and Stein studied the behaviour of R CrB at infrared wavelengths during the minimum of 1971–2 and found no change in the infrared component.[28] Lee also reported no change at infrared wavelengths during a minimum of RY Sgr.[29]

This contradiction forces us to rethink our models for the R CrB stars, for it cannot be the formation of graphite that causes the dips in the light curve. A model in which the star itself undergoes the fading is unattractive, but perhaps only because our current understanding of the nature of stars is too weak to accommodate such dramatic changes. We are therefore left with the alternative that the dust is always present and that occasionally and by devious means it is wafted into the line of sight. Such a hypothesis, using a second star to control the dust, was proposed by Humphreys and Ney.[13]

M DWARFS AND SUBDWARFS

A lot of work has been done on M dwarfs and subdwarfs. These are probably the most numerous stars in the galaxy. Being cool, their luminosities are best measured in the PbS region. This work is routine and rather unexciting because none of the stars have infra-red excesses. The nearest star in the sky, Proxima Centauri, is an M dwarf: the infrared measurements confirm that this M5e dwarf fits the calculated main sequence and has a temperature of 2,700°K.[30]

A related group of stars are the Haro-Chavira objects selected photographically because of their extreme redness. These were examined by Lee who found them to be normal cool stars with no excesses out to L.[31]

ε AURIGAE

There has probably been more confusion over the infrared excess in ε Aurigae than for any other star. As noted in Chapter 2 it began in 1964 when Mitchell claimed to have found a large excess at $10\mu m$. A few months later Low and Mitchell reported no excess in the star. Garbled versions of this were incorporated into the literature which followed. It was not until 1973 that Woolf published reliable magnitudes in the $5-18\mu m$ region which show that indeed there is an excess of a few tenths of a magnitude at these wavelengths.[32]

The interest in ε Aurigae is because of its peculiar eclipses. Every twenty-seven years something – we don't know what – passes in front of the F supergiant primary. The eclipse is partial yet the light curve is extremely flat bottomed. The star is not reddened, so optically thin dust cannot be responsible. At minimum the spectrum is still that of the F supergiant: there is no hint of the object which causes the eclipse. There is no secondary eclipse.

Many theories of the nature of the companion have been proposed and rejected; the one currently enjoying widest support is due to Huang.[33] In Huang's model the companion is a dense elongated cloud which is totally opaque. This stellar cigar orbits the F supergiant in such a way that every twenty-seven years it moves in front of and occults about three-quarters of the primary. The duration of the eclipse is dictated by the length of the cigar and hence the time it takes to cross the F star. But Huang's disc must radiate at least the amount of energy it absorbs from the F supergiant, and if, as seems necessary, it does so at a lower temperature than the primary, the ε Aur system must show an infrared excess. The small excess that Woolf found is just about right to be explained by Huang's cigar.

SYMBIOTIC AND VV CEPHEI STARS

In the grand scheme of matters celestial there are more types of object than man can conveniently comprehend. Doggedly he devises little boxes, each with a clearly written label, and endeavours to place every object he observes into one of these boxes. When, for example, it comes to spectral types he gives every star a tidy tag – A7 or G5 or whatever. This scheme would be fine were it not for the fact that some stars are confirmed misfits. The symbiotic stars are one group which quite simply do not conform. The name was coined by Paul Merrill to describe a type of spectrum which seemed to be a composite of a hot and a cool star. He borrowed the term from the botanists who use it to describe lichens – algae and fungi which live off one another in a harmonious arrangement.

Symbiotic stars have the molecular absorption-line spectra of late-type giants and supergiants. Superimposed on this is a rich emission spectrum of the sort produced by only the hottest of stars. A cool star could not possibly support the emission lines. But a hot star would dissociate the molecules responsible for the absorption lines. There is some evidence for a binary nature of many symbiotic stars, but somehow both characteristics may well be incorporated into a single star.

What happens in the infrared? Do symbiotic stars behave like hot or cool stars? All the accepted symbiotic stars have now been observed in the PbS region, and with a single exception they have the energy distributions of cool stars of about 3,000°K.[34,35] The exception is RX Pup which has a black body infrared excess. Nor do only the known symbiotic stars behave in this way: a number of emission-line stars once believed to be hot have turned out to have cool stellar continua in the infrared. These too are probably symbiotic.

Only a few symbiotic stars have been followed to 10 and 20μm. These generally show no infrared excess, including no silicate bump. Again the exception proves the rule, and in this case Z Andromedae, the prototype symbiotic star, is the exception. While behaving normally out to 10μm, at 18μm it is 3·5 magnitudes too bright.[32]

In 1974 Webster and Allen discovered a small group of southern hemisphere stars which appear to have the spectra of symbiotic stars but also have infrared excesses like RX Pup.[36] If these are confirmed to be symbiotic, RX Pup will no longer be the sole exception.

The VV Cephei stars in many ways resemble the symbiotic stars. They differ in having emission-line spectra of slightly cooler stars, but, more importantly, in being quite definitely binary. As for the symbiotic stars the infrared continuum is dominated by the cool star.[34] In addition it looks as though a little free-free radiation and a silicate bump is present in W Cep.[32] α Sco and R Aqr are two rather similar objects in which there is a late-type supergiant embedded in a nebula (probably of its own forming), and accompanied by a very hot star which excites the emission lines of the nebula. α Sco has only a tiny silicate excess despite being an oxygen-rich supergiant.[8]

Presumably, as for WY Gem and KN Cas, the hot star destroys the dust. In R Aqr the companion is not so luminous and there is a pronounced excess which resembles a silicate feature en route to becoming a black body.[24]

A star which is even more defiant of classification than the symbiotics lives under the unpretentious name of V1016 Cyg. Some years ago it was a contented M star, possibly a Mira variable. Then, quite suddenly, it brightened a thousandfold, lost all trace of its late-type heritage and adopted a very hot emission-line spectrum rather like that of a slow nova. Unlike a nova, however, it has shown no significant change since its eruption in 1967. The infrared continuum is that of a black-body dust cloud at about 750°K.[34] This dust varies in the infrared with a 450-day period in much the way that the stellar continua of Mira variables do.[37] Yet this infrared variation is not accompanied by any periodic change in the visible.

A star with a not dissimilar history is V1329 Cyg (formerly known as HBV 475). After its flare-up, V1329 Cyg had the spectrum of a Wolf-Rayet star with additional emission lines not normally associated with even these peculiar hot stars. Infrared photometry by Allen showed that the V1329 Cyg system also contains a cool star.[38]

GLOBULAR CLUSTERS

Globular clusters are aggregates of cool stars and have corresponding infrared colours. Two of them are bright enough to have been included in the IRC. M22 is one of these: the reason it appears redder than other globular clusters is not that it has an infrared excess but that there is some intervening dust in the line of sight which dims the optical cluster. Not far away is IRC − 20 385, a faint extended object at infrared wavelengths which was shown by Wray to be a heavily reddened globular cluster.[39] If the intervening dust were removed, IRC −20 385 would be one of the brightest globular clusters in the sky.

Globular clusters have not been studied much at 10μm. One which has is M15, and this was reported to have an infrared excess.[40] The observation is of low statistical weight and should be confirmed. If real, the excess presumably arises because a lot of the stars have thrown out dust clouds.

VARIABILITY IN THE INFRARED

Many cool stars are variables in the visible, and mention has already been made of variations at infrared wavelengths. As a general rule any late-type star which varies at optical wavelengths also varies in the PbS windows but with a much smaller amplitude. Variations at 10μm are generally tiny and hard to detect. Mira variables are well known to have amplitudes of as much as 10 magnitudes in the visible, but a variation of 2 magnitudes at K or L is rare indeed. At longer infrared wavelengths very little work has been done. Studies of the $2\cdot2\mu$m variability have been undertaken with the 62-inch survey telescope on Mount Wilson; $3\cdot5\mu$m measures were made by Don Strecker at Minnesota for his PhD thesis. Of the IRC stars, 12 per cent were found to be variable at K.

When we plot the energy distributions of Mira variables at maximum and minimum we find that the principal change has been in the temperature rather than the radius of the star. This, indeed, is apparent from the fact that the amplitude is greater in the visible, for if it were only the apparent size of the star which varied, all wavelengths would be equally affected. At minimum, Mira variables are cooler than at maximum light, typically by a few hundred °K. Once we have determined the temperature of a Mira variable we can estimate its size by assuming it to be a uniform black body. This is not really a good approximation since (a) the star will not be uniform but will probably have limb darkening and perhaps sunspots; (b) the fluxes at different wavebands are affected by absorption lines, and (c) we see to different depths into the star (and hence to different temperatures) at the various wavelengths. However, it is a good enough approximation from which to determine whether stars do in fact change size throughout their cycles. Indeed they do: Mira variables swell as they cool and shrink as they warm. It is the temperature variation which dominates in the visible because for the cooler stars (at temperatures below 3,000°K) the entire visible region lies on the very steep leading edge of the black-body function. Minimum light therefore occurs when the star is largest.

6

HOT STARS

In the preceding chapter we concerned ourselves principally with stars whose surface temperatures are less than about 5,000°K. Such stars emit most of their energy in the red or infrared part of the spectrum and are therefore a natural subject for infrared photometry. The stars to be described in the present chapter are hotter than 5,000°K, and most are considerably hotter. The peak of their energy distributions occurs in the visible or ultraviolet and throughout the infrared the stellar continuum emission falls rapidly. For this reason the IRC contains very few hot stars. Of its 5,612 entries, 40 are A stars, 27 B stars and only 3 are of spectral type O. There are also a few highly reddened O and B stars in the Cygnus OB2 association. At 10μm the situation is more extreme: the stellar continuum can be easily measured for only about 500 stars of spectral type A or earlier, and by 20μm this figure has slumped to under a score. It is apparent that unless hot stars have infrared excesses this chapter will be extremely short.

The reader who has thumbed through to the end of the chapter will therefore realise that some hot stars do have infrared excesses. We saw in the last chapter that infrared excesses are produced by circumstellar gas and dust. This applies whatever the temperature of the underlying star. When the star is particularly hot the optical spectrum contains immediate evidence of the presence of circumstellar gas: emission lines. Any stellar radiation shortward of 912 Å (the Lyman limit) is absorbed by hydrogen and ionises the atoms. Cool stars emit negligible quantities of this Lyman continuum radiation and can ionise very little hydrogen; hot stars are capable of ionising vast amounts. Ionised hydrogen radiates recombination lines, and in the optical region the strongest of these is the principal Balmer line Hα. If there is enough material around a hot star to give rise to an infrared excess there will be enough hydrogen to produce strong emission at Hα. In this chapter, therefore, we will consider only those early-type stars with Hα in emission. This restricts our

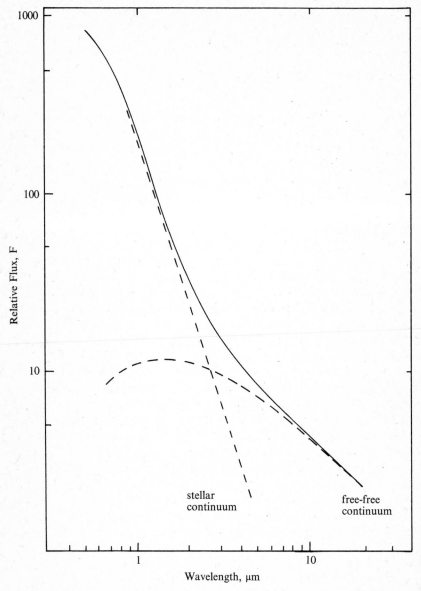

Fig. 20 The typical energy distribution of a hot star with free-free emission from its circumstellar gas.

sample to about 0·5 per cent of the recognised O, B and A stars plus a few thousand others of unspecified spectral type. We may divide these into two groups. The term *Be stars* is used as a general description of O, B, A and Wolf-Rayet stars with Hα emission, and this forms the first group; the second we call planetary nebulae. Empirically the planetary nebulae differ from the Be stars only in being hotter; evolutionarily they are quite distinct. There are no planetary nebulae and only a few Be stars in the IRC.

FREE-FREE EMISSION IN Be STARS

Most Be stars have infrared continua attributable to free-free radiation. This *Bremsstrahlung* occurs between the electrons and protons produced when the circumstellar hydrogen is ionised. Because ultraviolet photons do the ionising of the gas, its temperature is high: 10,000 °K is a typical figure. The λ^{-1} portion of the free-free curve (Fig 19) therefore extends on the short wavelength side right into the optical. The total power emitted by the free-free radiation cannot exceed the stellar flux shortward of 912 Å, and for Be stars this is a small fraction of the total output of the star. Free-free emission is therefore a small excess which first manifests itself around 1–2μm and becomes steadily more pronounced at longer wavelengths.[1] Fig 20 is a typical case. Free-free emission is easily detected in stars of spectral type B5 and earlier, but is very weak at later spectral types.[2]

FREE-FREE EMISSION IN PLANETARY NEBULAE

A Be star is essentially a hot star surrounded by a tiny dense nebula in which the hydrogen is ionised and from which free-free radiation and hydrogen recombination lines emanate. A planetary nebula is much the same: it is an extremely hot star surrounded by a large and rather tenuous nebula. The hydrogen in the nebula is ionised: it should emit free-free radiation. Moreover, because the central star of a planetary nebula may be more than five times as hot as a Be star, the amount of radiation in the Lyman continuum is considerably greater, and the energy radiated by the *Bremsstrahlung* (and by the optical emission lines) is correspondingly increased. In the optical the nebula can appear much brighter than the star. In the infrared the free-free radiation is many times stronger than the continuum of the central star. The energy distribution of planetary nebulae throughout the PbS region does indeed follow the λ^{-1} law.[3]

Because planetary nebulae are tenuous, the turnover of the free-

free slope from λ^{-1} to λ^{-3} occurs at very long wavelengths. The λ^{-1} portion observed in the PbS region can also be detected in the radio portion of the spectrum. By contrast Be stars, having denser circumstellar envelopes, become optically thick somewhere in the infrared and are generally too weak to be measurable radio sources. There is evidence that the energy distribution of Be stars becomes optically thick by 10μm,[1] and this corresponds to a number density of electrons (or protons) around 10^{11} cm^{-3}. The density of planetary nebulae range from 10^3 to 10^6 cm^{-3}.

DUST IN Be STARS

The majority of Be stars have insufficient circumstellar dust to give rise to an infrared excess over and above that of the free-free emission. This is not necessarily surprising: dust grains do not take kindly to ultraviolet radiation, nor to bombardment by protons. And there are prolific quantities of both in the immediate vicinity of Be stars. At greater distances from such stars, where these agents might have more diluted effect, there is very little gas and thus we cannot expect there to be much dust. The atmospheres of most Be stars do not extend far enough for dust grains to exist. But there are a few Be stars which have much more extended atmospheres. These are characterised by very rich emission spectra, and in particular by forbidden line emission.

Forbidden lines are so called because simple quantum theory predicts that they cannot occur. When an electron cascades down through an atom the emission lines it radiates are predetermined according to a set of quantum rules. From most of the energy levels it lands in there is at least one allowable escape route which the electron will take after pausing for about 10^{-8} seconds. Eventually it will land in the lowest allowable level, the ground state. But once in a while the electron lands in a metastable level – one from which the simple rules allow no escape. The electron is stuck on a shelf somewhere above the ground state. If this untidy situation arises in a laboratory the electron is not doomed to sit on its shelf forever. Within a very short time its parent atom will collide with another atom, and the collision shakes the electron from its metastable ledge. The energy freed from the electron is carried off by the colliding atom.

In gases more tenuous than can be produced in the laboratory, collisions are rare. Under these circumstances the quantum rules are slightly altered. After about 1 second the electron descends from its shelf emitting a forbidden line. Forbidden lines occur only in

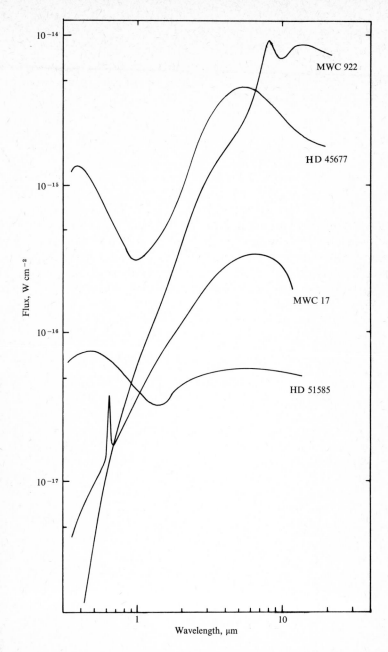

Fig. 21   Energy distributions of four early-type stars with circumstellar dust.
The spike in the visible continuum of MWC 17 is caused by very strong
Hα emission in this star.

tenuous gases where there are fewer than about 10^8 atoms per cm^3. Most of the emission lines in planetary nebulae are forbidden lines. In particular the so called nebulium lines which make planetary nebulae that characteristic green are transitions of [O III]. The square brackets signify that the lines are forbidden and O III is twice ionised oxygen.

Be stars, being cooler, do not ionise oxygen to so great an extent, and much of it is neutral. One of the common forbidden lines is therefore that of [O I]. [S II] and [Fe II] also frequently occur. There are about 100 Be stars with forbidden lines, and virtually every one has a big infrared excess resembling a black body. Fig 21 contains a few of the better-known examples.

It is interesting to detour into the history of this work. During the 1940s and 1950s when the field of Be stellar spectroscopy was opening up, many forbidden-line stars were studied by Swings, Struve and others. Most of these were symbiotic stars with clear evidence of molecular absorption bands: they had late-type affinities. In some forbidden-line stars no absorption lines were found and in one, HD 45677, the absorption lines are those of a hot B2 star. It was thought by some that HD 45677 should also be symbiotic, and possibly even a binary system – a B2 star and a late-type companion. Consequently in 1969 Pol Swings suggested to Frank Low that HD 45677 should be observed in the infrared in a search for evidence of the cool companion. The star was duly observed, and instead of a late-type component Low *et al* found a $600\,^{\circ}$K black-body excess.[4] This they attributed to an 'infrared star' orbiting the B2 primary.

The discovery of an infrared excess in HD 45677 prompted one of Low's co-authors, Susan Kleinmann (then Susan Geisel) to examine a number of forbidden-line stars at infrared wavelengths. She found most of them to have infrared excesses.[5]

Mrs Kleinmann attributed the excesses to dust forming in the extended atmospheres and noted that these stars are probably losing mass, like the cool supergiants, Swings (Jean Pierre – the son of the Swings mentioned above) and Allen analysed the spectrum of HD 45677 and could find no evidence for a binary nature to this star.[6] These two authors made a comprehensive survey of forbidden-line stars and found many to have infrared excesses. They also reversed the procedure and discovered forbidden lines in the spectra of emission-line stars with infrared excesses.[7] Infrared observations of nearly 250 stars were finally published.[2]

The brightest and best known of the forbidden-line objects is η Carinae (see plate 4). Its infrared excess is the most extreme

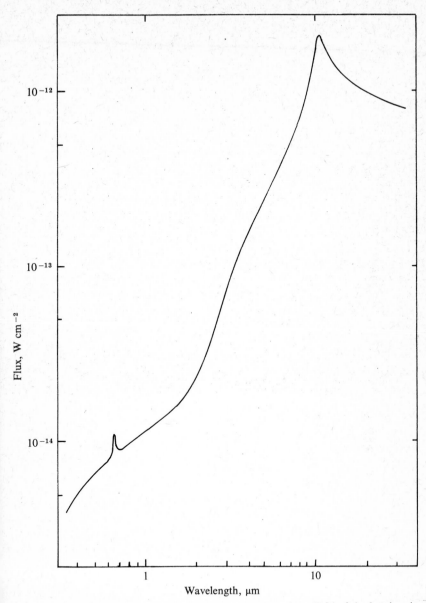

Fig. 22   The energy distribution of η Carinae. More than 99% of the luminosity of this object is radiated in the infrared; the total luminosity is about as great now as was the optical emission at the peak of its nova-like outburst. η Car is the most luminous single object known: the visible and infrared luminosity exceeds one million times that of the sun.

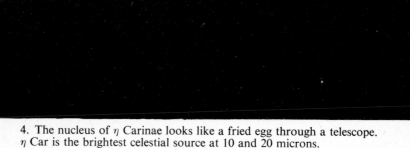

4. The nucleus of η Carinae looks like a fried egg through a telescope. η Car is the brightest celestial source at 10 and 20 microns.

discovered (Fig 22) and was known as long ago as 1968.[8] Because of its unique nature – extreme luminosity, nova-like variations, nebulous appearance – it was not at first likened to the other forbidden-line stars. But more recently several stars have been found with infrared excesses and optical spectra remarkably similar to η Car.[9,10] In each case it was the discovery of the infrared excess which prompted the subsequent optical spectroscopy. Four of these stars were discovered to emit Hα during the Mount Wilson survey and are named MWC (Mount Wilson Catalogue) 645, 819, 922 and 939. Others have been classified as planetary nebulae.

The infrared excesses in the forbidden-line stars are roughly black body in nature, as an inspection of Fig 19 will reveal. There is no indication of silicate bumps and the nature of the emitting dust is not known. The sole exception to this is η Car which does have a modest silicate excess superimposed on its black-body continuum.[11] The only object of this kind which has been resolved in the infrared is again η Car: it is about 10" arc across at 10 μm and much smaller at 2 μm.

Although the infrared energy distributions of several forbidden-line stars approximate quite closely to that of η Car, none is quite so similar as that of IRC+10 420, 'Eta Junior'.[12] IRC+10 420 is not an emission-line star but a fairly normal F supergiant, and what is really remarkable is that its spectrum closely resembles that of η Car in 1892, about fifty years after its nova-like outburst and a few years before it developed its present emission-line spectrum. IRC+10 420, however, has no history of nova-like activity to its credit, so how similar the two objects are is hard to say.

FREE-FREE, MASS LOSS OR COOL COMPANIONS?

While dust is the natural and conventional mechanism to explain the infrared excesses in Be stars, the theoretical difficulty of forming and maintaining it in so hostile an environment has troubled some authors and caused them to seek alternative explanations. One attractive proposal by Dyck and Milkey involved free-free radiation of the form found in the NML Tau stars.[13] Dyck and Milkey calculated that *Bremsstrahlung* could occur from the zone outside the circumstellar H II region where hydrogen is neutral but heavier atoms are ionised. The electrons are generated by the heavier ions and *Bremsstrahlung* occurs around the hydrogen atoms. The low temperature of the gas in this region causes the free-free energy distribution to rise in the PbS region instead of the optical. Dyck and Milkey constructed models which fitted the observations well and required no dust. Their model has been rejected now on two grounds. First, the expected free-free component is much smaller than the observed excesses, and secondly the 1 μm peak predicted for this type of free-free is not observed.[14]

Humphreys and Ney revived Low's original suggestion that the forbidden-line stars are all binaries and that the infrared excesses are no more than NML Cyg stars orbiting the B primary.[15] This explanation cannot work for some forbidden-line stars, in particular η Car, and therefore is probably not the correct one for the remainder.

So we must fall back on the dust. A piece of evidence which partially supports the dust hypothesis is the reddening of the stars. The forbidden-line stars are reddened by dust which must lie somewhere in the line of sight. The energy taken out of the star by this reddening is in most cases balanced by the energy radiated in the infrared component. And this is just what we would expect if the dust lies in a shell around the star. Unfortunately, in some cases the reddening of the star is insufficient to account for the infrared excess. HD 45677 is one such example. We must assume that the dust is not uniformly distributed around this star but lies in clumpy clouds. The infrared signal averages out the clumps, but the star may be seen through a clearer portion and hence not appear very reddened. A model for hot stars which seems to be consistent with all the known facts has the dust forming in dense neutral clouds within the ionised region. In these clouds the effects of ultraviolet radiation and proton collisions are not felt.[16]

DUST IN PLANETARY NEBULAE

The Be stars generally have dust at temperatures around 1,000 °K, and this produces an excess at wavelengths as short as 1·5–2 μm. In planetary nebulae the dust is much cooler and may give rise to no black-body emission shortward of 8 μm. Energy distributions of some representative planetary nebulae are given in Fig 23. The most comprehensive lists of infrared photometry of planetary nebulae are those of references 17 and 18.

Silicates appear to be absent from planetary nebulae. Rising energy distributions in the 10–20 μm region are common, but whether this is an emissivity effect or is caused by cool dust we cannot yet say. Infrared excesses have been found in about half of the planetary nebulae sampled, and this figure would probably be greatly increased if our detection limits could be improved. It is not impossible that all planetary nebulae contain a certain amount of dust. The dust appears to be distributed roughly like the gas, ie as a hollow sphere or ellipsoid centred on the star. In some very extended nebulae the dust is more concentrated towards the centre. There is some correlation between the infrared excess and the type of central star, but no clear pattern has yet emerged.

The extended planetary nebulae probably evolve from late-type stars. Current theories of their formation begin with cool supergiant stars. We have seen that mass loss is prevalent in such stars, and if the star passes through a phase in which the rate of mass loss increases dramatically, most of the outer layers of the star can be shed into a roughly spherical expanding cloud leaving only the core in the middle. The core of a cool star is very hot – perhaps 50,000 °K – and this ionises the expanding shell to produce a planetary nebulae.

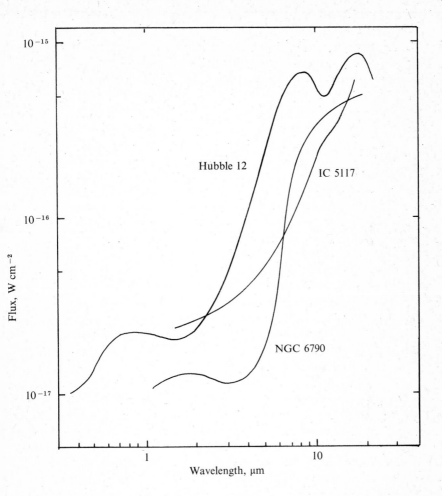

Fig. 23 The energy distributions of three planetary nebulae. The dust is cooler than in normal stars and the infrared continua are shifted correspondingly to the right.

As the gas expands it cools sufficiently for dust to form. But here we have a problem. According to the models, oxygen-rich stars can undergo this transformation, and the dust which condenses from their outer layers should be silicates. Where are the planetary nebulae with silicate bumps? At the moment we do not know.

Not all planetary nebulae are extended, green and planet-like. About 40 per cent of the objects currently called planetary nebulae are in fact stellar, or at least have not been resolved on photographs. Why then are they called planetary nebulae?

When we examine the spectra of planetary nebulae we find a very characteristic set of bright emission lines that recur time and again. Foremost among these are the lines of the Balmer series of hydrogen, helium either neutral or singly ionised, the 'nebulium' lines of [0 III], the violet lines of singly ionised oxygen [0 II], and in the red [N II]. Because the star is very hot the emission lines are strong and in many planetary nebulae the stellar continuum is relatively weak, so that at first glance only the emission lines are seen.

Spectroscopic surveys, made by placing a large prism in front of a refracting telescope or Schmidt camera, so that every star is spread into a tiny spectrum on the photographic plate, reveal the existence of stellar objects with the same spectral features. It is therefore quite reasonable to call these planetary nebulae, and logical to assume that they are younger specimens which have not expanded to a resolvable size. From the line intensities we can tell that most of them are denser than the extended nebulae.

Now consider a fainter object. Many of its emission lines will be too weak to be detected by an objective prism survey: maybe only the strongest, Hα, will be visible. There are many types of object besides planetary nebulae which have Hα in emission; to name a few: Be stars, forbidden-line stars, H II regions, T Tau stars, some M stars, symbiotic stars, emission galaxies. To distinguish planetary nebulae from this selection is an almost impossible task. Only one criterion can be adopted: if there is a stellar continuum present on the plate the object is a Be star or similar object; if not it is a planetary nebula.

The result of this inadequate selection mechanism is that planetary nebula catalogues are overburdened with objects which are not planetary nebulae, and among the stellar nebulae this problem is most acute. Probably half of the so-called stellar planetary nebulae are not really planetary nebulae at all if we use a strict spectroscopic definition. But these stars are faint, and to get spectra of them involves a considerable amount of telescope time. Here infrared photometry can help. In a few minutes' observation it is possible to identify a stellar planetary nebula with one of three classes according to whether there is or is not a 2·2 μm signal and whether the colours are those of a late-type star or a dust excess. The undetected sources are probably planetary nebulae or simple Be stars; the late-type stars are probably symbiotic and the stars with dust clouds are mostly forbidden-line stars. Most known stellar planetary nebulae have been thus examined by Allen.[10,19] From a sample of 372 stellar planetary nebulae, 79 have late-type continua and 43 have dust excesses.

M2–9, Mz 3 AND THE FORBIDDEN-LINE STARS

Two planetary nebulae which deserve particular mention are M2–9, discovered by Minkowski in 1946, and Menzel 3. Plate 5 shows

5. M2-9. This strange symmetrical nebula is an infrared source but is very weak at radio wavelengths.

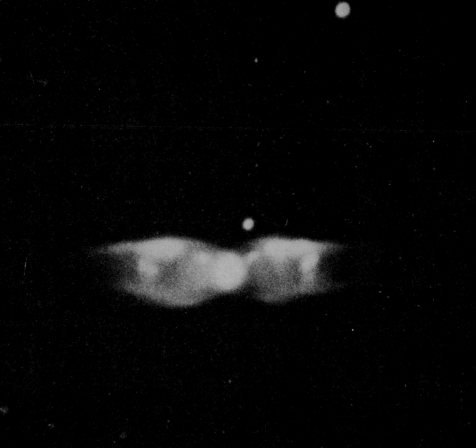

M2–9 and it may be seen to resemble a butterfly with sharp-edged symmetrical wings. Mz 3 is essentially its twin except that the wings appear to be foreshortened. M2–9 was found to radiate strongly in the infrared by Allen and Swings: more than 90 per cent of its luminosity is in the cool dust radiation.[20] Mz 3 is again similar in this respect.[21]

Amongst extended planetary nebulae so intense an infrared excess at so high a dust temperature (over 700 °K) is rare indeed. Considering also the unusual shape, which is found in no other known planetary nebula, we begin to suspect that M2–9 and Mz 3 are not normal planetary nebulae at all. This suspicion is further strengthened by their spectra. Both objects combine features found in the spectra of planetary nebulae and of forbidden-line stars. The nebular emission lines of [O III] and [N II] are strong, but so are the [Fe II] and [Fe III] lines typical of the forbidden-line stars. It is tempting to suggest that, unlike conventional planetary nebulae, M2–9 and Mz 3 were formed by luminous B stars losing mass in a steady stream, and that the gas and dust shed in this process was funnelled into narrow jets on either side of the star by the effects of magnetic fields. This is, however, no more than speculation on the part of the author.

FG SGE

I will take the liberty of describing a planetary nebula with no known infrared excess. This is the variable star FG Sge which lies at the centre of an annular planetary nebula discovered by Henize and labelled He 1–5. The nebula is some 30″ arc across, and presumably radiates a small amount of free-free radiation which has yet to be detected. The star, on the other hand, is easy to measure in the PbS region. There is no excess: the infrared flux comes from the continuum. In the infrared the star is brightening while at visible wavelengths it is almost constant.[2] Thus its colour temperature is falling. Accompanying this slow change in temperature is a corresponding shift in the spectral type from a late B supergiant in 1962 to a G supergiant in 1971. At no known time has its spectrum resembled in any way the stars one usually finds at the centres of planetary nebulae. FG Sge is a most unusual and, sadly, a rather neglected star which should be monitored over the next few decades: here we may be witnessing the death of a star.

The Wolf–Rayet stars are strange creatures. They are very hot, very luminous objects which probably belong near the top end of the main sequence, above O and B stars. But their spectra are quite unlike those of main-sequence stars, for they comprise a collection of extremely broad emission lines. So broad are these lines that they overlap considerably and essentially obliterate the underlying continuum of the star. The emission lines originate in a turbulent atmosphere which is being thrown off by the star at speeds that may exceed 1,000 km/sec. This is comparable with the ejection velocities of novae, but whereas novae are content to puff off a single shell of material, wolf-Rayet stars are doing it all the time!

Spectroscopically, the Wolf–Rayet stars comprise two parallel types, the carbon and nitrogen sequences. WC stars have emission lines principally of hydrogen and helium and C III and C IV. In WN stars the carbon is replaced by N III and N IV, and a few oxygen lines. The two sequences are each subdivided into five types, numbered from 5 to 9, such that stars in higher-number classes have lower excitation spectra (they are thus probably cooler) and smaller ejection velocities. The reason for the two distinct types is not clear. Some think that it is simply an abundance difference, analagous to carbon- and oxygen-rich late-type stars. Others are not so attracted to the idea that such great abundance variations exist among early-type stars and argue that differing conditions in the atmospheres of Wolf–Rayet stars favour the excitation of carbon lines in one set, nitrogen in the other.

Now, mass loss is something we have come to associate with infrared excesses, but in the case of the Wolf–Rayet stars our optimism must be tempered by the fact that dust grains are not very robust; the hot and violent conditions around Wolf–Rayet stars are hardly conducive to grains leading a peaceful and contented existence. The observational situation is both interesting and unclear.[22]

The principal infrared continuum found in Wolf–Rayet stars is free-free radiation, and this is no surprise. What is surprising is that the 2 μm colours of WC stars are systematically redder than those of WN stars. The H–K colour index for WN stars ranges from $-0^m\cdot1$ to $0^m\cdot35$; with only two exceptions in WC stars the range is from $0^m\cdot35$ to $1^m\cdot7$. So distinct are the two groups that infrared photometry alone could be used to spectral type Wolf–Rayet stars into the two classes. There are four possible reasons for this difference in colour:

(i) The free-free component in WC stars could be stronger than in WN. Since in the WN stars the free-free continuum is already pretty extreme, it seems rather unlikely that it could be much stronger in the WCs.

(*ii*) The WC stars have extra emission lines in the K window. This too seems unlikely for no suitable strong emission lines are known in the relevant waveband. Besides, observations of the brighter stars at other infrared wavelengths fit smoothly with the H and K data. Nor is it likely that WN stars have extra lines at H making them appear bluer.

(*iii*) The underlying continuum of WC stars is redder than that of WN stars in the 2 μm region. No obvious reason for this is known.

(*iv*) The WC stars have small dust excesses added to their free-free continua. A very weak black-body dust continuum is indistinguishable from a free-free continuum except at the longest wavelengths, and there the photometry is insufficiently reliable. Therefore, infrared photometry cannot decide whether dust is indeed present in all the WC stars. Two pieces of evidence do, however, suggest that this may be the correct explanation.

First there are the two exceptions, γ^2 Velorum and θ Muscae, WC stars with H–K colours of WN stars. Whatever causes the extra excess in other WCs is not present in these two stars. The only other known difference between these and other WCs is that they have very luminous O star companions. Many WC stars have B companions which are less luminous. An obvious possible effect of the luminous O stars is to destroy or disperse any dust that is trying to form.

Second, and more important, some of the WC stars do have very large H–K colours, and photometry of these at longer infrared wavelengths delineates the excess very clearly, showing it to be black-body in nature. The most extreme of these is Ve2–45, discovered by the Belgian astronomer Velghe. It is a 13-magnitude star in Sagittarius, but its dust excess is so pronounced that it was recorded in the IRC, and at 20 μm its magnitude is −2·6. All the WR stars with large and obvious dust excesses are of spectral type WC 9 – ie they are the coolest and least violent of their kind – just the stars we most expect to be able to support circumstellar dust clouds.

The WC spectral sequence was recently extended to WC 10 by the inclusion of four stars which seemed to fit naturally onto the end. Webster and Glass described these four:[23] they have lines of C II and C III, of lower ionisation than traditional WC stars, and the lines are narrower. Two of them, M4–18 and Henize 1044, were originally classified as stellar planetary nebulae and one, V348 Sgr, has been described as an R CrB variable star. The fourth is CD $-56°8032$. Perhaps not surprisingly these WC 10 stars have prominent infrared excesses. Indeed CD $-56°8032$ has the second largest $K-L$ colour index of any emission-line star, exceeding even η Car by half a magnitude.

Wolf–Rayet stars also occur as the central stars of planetary nebulae. This is rather worrying because Wolf–Rayet stars are believed to be young objects whereas planetary nebulae are old. It seems that the two types – the field and the planetary nebula Wolf–Rayets – are not related. Rather, Wolf–Rayet spectra can be produced by any star of suitable temperature which is losing mass fast enough. Some planetary nebula nuclei pass through the appropriate phase.

Curiously, the Wolf–Rayet stars in planetary nebulae also divide neatly into the carbon and nitrogen sequences, although WN is rare. More curiously, planetary nebulae with WC central stars frequently have dust excesses in the PbS region; those with WN nuclei do not.

The infrared data seem to be trying to tell us that WC stars throw off gas from which some sort of black-body dust can form, whereas WN stars eject no material conducive to black-body dust growth. And the logical interpretation of that would be that WC stars are indeed carbon-rich and WN stars are not.

NOVAE AND SUPERNOVAE

The most obvious examples of mass loss are those provided by novae and supernovae. But are the conditions around novae too extreme for dust to form? The first nova to be studied at infrared wavelengths was Nova Ser 1970.[24] Its behaviour has subsequently been mimicked by other novae and we may take it to be normal.

When a nova explodes, a small portion of the outer layers of the parent star is shed in a single hiccup. A thin shell of dense material is thrown out by the star and this expands more or less uniformly, cooling as it does so. There comes a point where the temperature is low enough for dust to condense in the shell, and when that happens the nova develops a black-body infrared excess over a period of probably a few days or less. This is typically about ten days after maximum light. The dust moves away from the star with the gas, and as it does so it cools. Over the next few months, therefore, the colour temperature of the infrared component falls until finally the dust becomes too cool to be observed.

113

It might be expected that a similar sequence of events occurs in supernovae. This is not the case. The bright supernova in NGC 5253 (in 1972) afforded an excellent test object, for the expected continuum was bright enough to be measured in the PbS region. No infrared excess was observed;[25] nor has any been found in any other supernova. It seems that here, at last, we have found somewhere totally lethal to that almost ubiquitous phenomenon of circumstellar dust.

THE CRAB NEBULA

If supernovae are unproductive, what about their remnants? The biggest and brightest supernova remnant is, of course, the Crab Nebula, M1 (often mistakenly described as a planetary nebula). Infrared observations of the Crab Nebula have been tackled by many observers, and measurements to L[8,26] and M[27] were made as long ago as 1968. The Crab finally yielded at 10 μm in 1971.[28] The observations out to 5 μm are entirely consistent with the synchrotron radiation mechanism which links the optical and radio regions. The 10 μm data suggest that there is an excess, but the signal/noise ratio of the measurement is frighteningly low and the observation may be proved spurious.

The actual stellar remnant of the 1054 explosion is the pulsar NP 0532. This is the only pulsing radio source to have been detected at optical wavelengths, and is an obvious candidate for an infrared astronomer with a sufficiently large telescope – ie a 200-inch. The CalTech team have indeed measured NP 0532 out to 3·5 μm, and at K they were able to examine the light curve (or, to use the radio-astronomers' terminology, the pulse profile).[29] At best they could tell the 2·2 μm light curve is identical to that in the optical. There is no infrared excess at K or L.

X-RAY STARS

In our tour of the hotter types of stars we must terminate with the X-ray stars, at least one of which (Sco X-1) is probably one of the hottest stars known. Several stars identified with galactic X-ray sources are bright enough to be measured out to K or L, and a number of papers have been published describing them. In chronological order these are: Sco X-1,[30] HD 15391[31,32] and HD 77581.[32,33] In the two HD stars the X-ray emission probably comes from an unseen companion. None of these stars shows any peculiarities beyond the occasional free-free excess at any of the wavelengths employed.

An interesting observation of Cyg X-3 was made during its 1972 radio flare. This source lies behind the dense Cygnus Rift, and no optical identification has been made. But an infrared source was found very close to the radio position, and the colours of this source could be explained by a hot star considerably reddened by interstellar dust.[34] Further observations showed that the infrared source varies with a 4·8 hour period exactly as does the X-ray emission.[35] In addition, the 2·2 μm observation showed occasional flares in which the source brightened by up to a magnitude for about one minute. Cyg X-3 is though to be a similar object to Sco X-1 and it is unfortunate for the optical observers that the Cygnus Rift intervenes and prevents us seeing it.

INTERSTELLAR EXTINCTION

We have seen that the only normal stars which do not appear to have infrared excesses are the non-emission B stars. Only for these stars can we predict to within a few percent the continuum in the infrared. This is an important thing to do if we wish to study the wavelength dependence of interstellar extinction.

All predicted reddening laws have negligible extinction at 10 μm. Even if a star has 10 magnitudes of extinction in the visible, its 10 μm magnitude should be cut down by less than 0·1 magnitudes. A measurement of the star at 10 μm therefore provides a very reliable estimate of what magnitude the star would have at any other wavelength in the absence of interstellar dust. Hence the dependence of extinction on wavelength, the reddening law, is deduced and from it we can gain some inkling of what types of material lurk in the great unknown reaches of the galaxy.

This approach was employed by Johnson, and the best summary of his work will be found at reference 36. A few of Johnson's stars had emission lines, and some of these appeared much brighter at 10 μm than at the shorter wavelengths. There is therefore some suspicion that not all his reddening curves are reliable but that some were distorted by the presence of infrared excesses. This, however, does not detract from the bulk of Johnson's work. The reddening curves he found in some parts of the sky agree very closely with theoretically derived laws. It is disappointing that the interstellar smog produces no identifiable features in the optical and infrared portions of the spectrum; at least none that Johnson was able to identify. We shall delve more deeply into this question in Chapter 9.

7

YOUNG OBJECTS

The stars that we have considered so far are mostly in a fairly advanced phase of evolution. The late-type giants and supergiants of Chapter 5 are produced by stars evolving off the main sequence; their mass loss is induced by a change in their internal combustion processes. Planetary nebulae are at an even later stage in the story. Novae and supernovae are more violent manifestations of the same effect. Be and forbidden-line stars are also probably beginning their evolution off the main sequence. In their case the mass loss may be produced, or at least initiated, by an overly rapid rotation: by centrifuging off material, Be stars are hastening their evolution, and they may have only recently settled onto the main sequence. With the possible exception of the enigmatic Wolf-Rayet stars, none of the stars we have discussed are very young, and this situation is remedied in the present chapter.

When a star forms, it does so as a local density excess in a dust and gas cloud. The inner regions of the knot grow denser and denser until at length nuclear burning is triggered and a star is born. At that instant the pattern of the cloud is disturbed. No longer can material keep falling unimpeded towards the centre under the influence of gravity. A new force, radiation pressure, is introduced into the system: the newly created photons hurrying on their outward pilgrimage bump into the inward-migrating dust grains, halt them and even reverse their motion. Seen in this context a star is a subtle invention of Nature to avoid that embarrassing singularity in the laws of physics we call the black hole. More accurately it is Nature's delaying tactic, for it seems that once the nuclear-burning phase has ended, the stellar material must once again precipitate itself steadily closer to the gravitational plughole of its event horizon. Until, that is, some other phenomenon we have yet to identify takes control. But I digress.

New-born stars should be enveloped in dust clouds; these they will disperse long before they reach the main sequence. In evolving to the main sequence, however, most stars probably find it necessary to adjust their weights slightly by throwing off a little extra material, and this has the result of prolonging their dust-cloud phase perhaps until just after they have taken on the adolescent responsibilities of the main sequence. Thus we expect young stars to radiate strongly in the infrared because of the dust that surrounds them; in this we are not disappointed.

It is not always easy to recognise a star which is approaching the main sequence. The sort of optical properties which usually characterise this phase of evolution are: association with nebulosity, especially H II regions; irregular variability; emission-line spectra. The biggest classes of recognised young objects are the T Tau stars and the Orion variables. These differ only in minor respects, eg the form of their light curves, and for our present purposes we can regard them as the same. There are several hundred of them known, mostly as a result of the work of George Herbig of the Lick Observatory. The majority of the T Tau stars lie in the dark clouds that throng the constellation of Taurus, while the Orion variables are associated with the emission nebulae in Orion. The Taurus and Orion clouds belong to the same spiral arm and may be connected as an elongated feature like the Cygnus Rift. The principal difference between them is that the Orion material is ionised by the very hot stars which have formed there; in Taurus only cooler stars are forming.

Smaller concentrations of the same types of young stars occur in other portions of the spiral arms. Cygnus contains a fair number, in particular in the vicinity of the Pelican and North America Nebulae. Ophiuchus and Scorpio contain others.

Young stars are also found in many galactic clusters, in association with brighter and more massive stars which have lately reached the main sequence. The cluster of stars at the centre of the Orion Nebula is one such formation, but the best known example is NGC 2264 in Monoceros, the open nebulous cluster which contains the 5th magnitude O star S Mon.

Finally, a number of nebulous emission-line stars seem to exist in isolation or in very small groups. These too are young, and most of them are of fairly early spectral type, being Ae or Be stars.

THE INFRARED OBSERVATIONS

Historically, R Mon was the first young object found to have an infrared excess, by the Mexican astronomer Mendoza in 1966.[1] The

6. R Mon is the semi-stellar object at the apex of Hubble's variable nebula NGC 2261. This was one of the first objects found to have a prominent infrared excess.

energy distribution of R Mon is included in Fig 24, together with some T Tauri stars. The object itself is particularly interesting. In plate 6 R Mon is at the tip of NGC 2261, Hubble's variable nebula. Many photographs have been taken of R Mon – it was, in fact, the first object to be photographed with the Palomar 200-inch telescope – and none has shown the object to be a star. On the best photographs R Mon is a tiny nebula about 1″ arc across. Of course,

Fig. 24   Three T Tauri stars with infrared excesses. Note the strong Hα emission in R Mon.

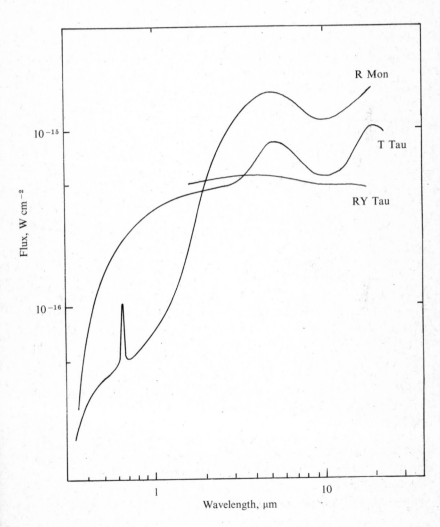

there almost certainly is a star somewhere in the middle, but it must be surrounded by a very dense dust and gas shroud. It therefore comes as no surprise that more than 95 per cent of the energy of R Mon is emitted by the dust, which has a colour temperature of about 750 °K.

T Tauri also lies in association with a variable nebula, this time Hind's nebula NGC 1554–5. It too was found by Mendoza to have an infrared excess, as were a number of other T Tau stars.[1,2] Energy distributions of the two best-known T Tau stars Mendoza investigated are included in Fig 24.

Mendoza's work was not followed up for several years. Indeed the entire field of the infrared emission of young stars has been much neglected by infrared astronomers until it was revived in the early 1970s by Steve and Karen Strom and Martin Cohen. The Stroms published infrared photometry of the NGC 2264 cluster in 1972, and although most of their photometry must be rejected because of the low statistical significance of the data, the basic finding of their research stands – namely that the young stars in this cluster have infrared excesses due to circumstellar dust.[3] One star in NGC 2264 deserves particular mention: it is variously known as Walker 90 and L Hα 25. The latter nomenclature is for the Lick survey of stars with Hα in emission, made by Herbig. The abbreviation was changed to Lk Hα for later discoveries in this survey, some of which we shall meet below.

L Hα 25 has a pronounced infrared excess – about 90 per cent of the energy being radiated by the dust. By rights, therefore, L Hα 25 should be considerably obscured and reddened at optical wavelengths by its circumstellar dust clouds. In fact, the reddening is negligible. Now, we could make the usual arguments, as in the last chapter, about seeing the star through a gap between its dust clouds so that there is no absorption and no reddening in the visible. That would be fine but for one detail. We know just how bright L Hα 25 should be, for it is a member of NGC 2264. It is two magnitudes fainter than other stars of the same spectral type. In other words the extinction is present in about the right amount, but the reddening is not. Is this an example of a peculiar reddening law? We cannot be certain. If it is, the interpretation must be that the dust grains are all much bigger than those around other stars, for only when the grains (or very dense clumps of grains) are considerably bigger than the wavelength of the relevant radiation (in this case about 0·6 μm) does the extinction affect all wavelengths equally. But why should the smaller dust grains be absent? Where did they go?

The Stroms have subsequently examined other young clusters finding, as for NGC 2264, that the young emission-line stars have circumstellar dust excesses.[4] The immense task of observing the Orion (M 42) cluster at infrared wavelengths was tackled by Penston,[5] and some of the brighter stars were observed by Ney, Strecker and Gehrz.[6] From these two papers we may deduce two features which

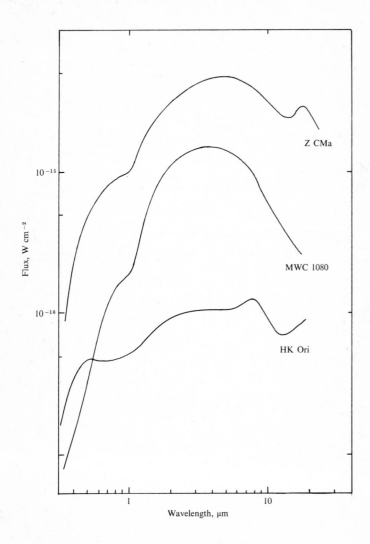

Fig. 25   Three young hot stars with infrared excesses.

seem to hold true for the other young clusters too. First, infrared excesses are correlated with optical variability: this suggests that the irregular variations of young stars are produced by the swirling in-homogeneous dust clouds that orbit them. Secondly, the hotter stars have excesses only at the longer wavelengths. This result – that the hotter the star the cooler the dust – seems to apply equally to forbidden-line stars and planetary nebulae.[7]

Stars in young clusters were also studied by Cohen for his Ph D thesis and published in one of a three-part pot pourri of infrared observations in 1973.[8] Cohen devoted another portion of his thesis to the T Tau stars which had been so neglected since Mendoza's time: he made extensive observations at longer wavelengths, some of which are incorporated in Fig 22. He sought evidence of variability in the infrared, but found nothing this author regards as conclusive because of the unreliability of 10 μm photometry. Nor did he find evidence of silicates in these stars. A more comprehensive survey has recently been completed by Glass and Penston who sampled eighty-nine variable stars thought to be young.[9] Their PbS photometry revealed infrared excesses in many of these stars, mostly those of spectral type A or F. They identified a few highly reddened stars too.

The young, isolated Ae and Be stars have massive infrared excesses. Some of these were first discovered by Gillett and Stein[10] and further investigated by the Stroms,[11] Cohen[8] and Allen.[12] Fig 25 contains energy distributions of a selection of them, and it will be seen that some are quite heavily reddened by their circumstellar dust clouds. Again there are no silicate bumps. The nebulae associated with the young Ae and Be stars are themselves interesting. Many of them are of the form described as *cometary*, and perhaps the best examples of this type of nebula are NGC 2261 (R Mon) and Lk Hα 208 (see plate 7). Cometary nebulae have soft edges in contrast to the sharply defined wings of M2–9 and Mz 3 discussed in the last chapter. Cohen has found that the stars in the cometary nebulae have an extra excess at 20 μm over and above the principal black-body excess fitted through the shorter wavelengths. This presumably means that in addition to the dust clouds causing the short-wavelength radiation, there are denser clouds at greater distances from the stars where the equilibrium temperature is much lower. Cohen thinks these may lie in flattened discs, which by absorbing light in one plane produce the cometary shape of the nebulae.

None of the stars mentioned so far in this chapter was bright enough to be included in the IRC, although Z CMa (see plate 8 and MWC 297 fell short by less than half a magnitude. At 2·2

7. The infrared source Lk Hα 208 is
embedded in a classical egg-timer nebula.

8. Z CMa is an irregular variable star distinguished by the small nebula to its north.

μm the brightest of these stars, Lk Hα 101, is magnitude 3·1 and just scraped into the IRC as entry +40 091 because of a small additional free-free contribution from the nebula NGC 1579 in which it is immersed. At visible wavelengths this star is 17th magnitude: its infrared excess therefore looks pretty impressive as Fig 26 shows.

Although an IRC entry, Lk Hα 101 was not recognised by the CalTech team. The identification resulted when Cohen plotted IRC source positions on the Palomar Observatory Sky Survey prints to seek optical counterparts.[13] Photometry was published in 1971.[14] Cohen also recorded a number of nebulous stars in his study of IRC sources.[15] Some of these may be young objects, but they have not been further investigated.

VARIABILITY

A feature of most young stars is their optical variability which is always of an irregular nature. It is not impossible that all this variation is caused by the swirling dust clouds which pass in front of the star. On the other hand contracting stars are likely to be unstable and intrinsically variable. Probably both mechanisms contribute. A possible way of deciding which is more important is to look for correlations between the amplitude of a variable and the strength of its infrared emission: if the dust causes the variation, the stars with little of it should vary least. Insufficient data have yet accumulated to say whether the correlation is valid, although as noted the existing data suggest that it is (eg reference 5).

FU ORI AND V1057 CYG

On old photographs FU Ori was a 16–17 magnitude star, probably T Tau in nature. In 1936 it underwent a dramatic rise in brightness to 9th magnitude, and in the subsequent four decades it has dimmed only a little. The behaviour, then, resembles that of V1016 Cyg and V1329 Cyg. But the spectrum after its outburst was not the rich emission-line spectrum shown by the two Cygnus stars. Rather it was, and still is, that of an F2 star with a few weak emission lines.

When FU Ori undertook its brightening, it changed from a pre-main-sequence T Tau star to what seems to be a young F star which has probably just reached the main sequence. Like many young stars there is a small nebula attached, in this case a reflection nebula. If FU Ori typifies the normal behaviour of a star attaining the main sequence, our theories of this phase of evolution are in need of

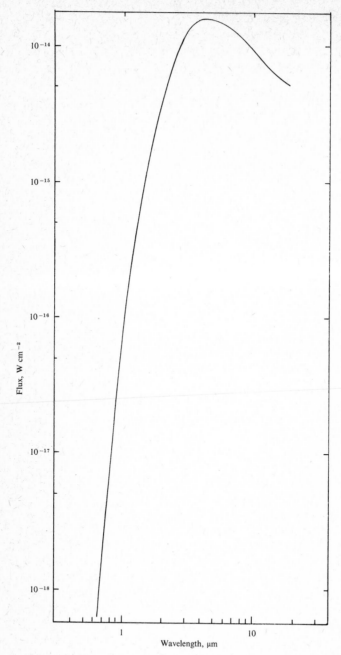

Fig. 26   Lk Hα 101 has a very extreme infrared excess.

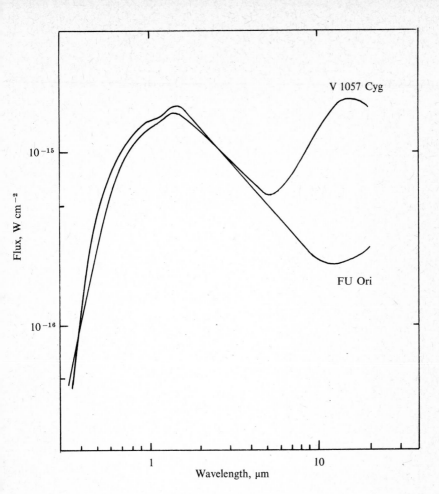

Fig. 27 FU Ori and V1057 Cyg. Both stars have undergone abrupt increases in luminosity.

revision, for none predict a sudden thousandfold increase in visible brightness. For this reason FU Ori was long considered to be an exceptional case.

But in 1970 another star did the same thing. This star was originally known as Lk Hα 269, and was a faint T Tau star in the North America Nebula. It is now a 9th magnitude A7e star illuminating a wispy reflection nebula, and is usually called V1057 Cyg.

Both FU Ori and V1057 Cyg have infrared excesses (Fig 27).[14] At the short wavelength end it is difficult to decide whether the excess is caused by free-free emission or dust; at longer wavelengths we can say with greater confidence that dust is responsible. The nature of the infrared continuum in V1057 Cyg in particular has been the subject of several papers.[14,16]

THE ROLE OF FREE-FREE RADIATION

In the last chapter I raised the question of whether the infrared excesses in hot stars could be explained by free-free radiation from a cool electron gas. I came down against this explanation and in favour of the traditional dust model. The same question can be asked about the stars of this chapter, in particular the young Ae and Be stars. As before, my preference is for the dust model which seems the natural one; but it should be made clear that the free-free hypothesis is not necessarily ruled out. For a review of the infrared excesses in young stars in the light of the free-free model the reader is referred to Strom's review paper at reference 17.

HERBIG-HARO OBJECTS

Not all the objects in young clusters, and in particular in the Orion Nebula region, are stellar. A class of purely nebulous emission-line objects was recognised independently in the 1950s by Herbig and Haro: these are now known as Herbig-Haro objects, and it is perhaps worth pointing out that since Haro is a Mexican his H is not pronounced.

On photographs Herbig–Haro objects rather resemble jellyfish. They contain bright knots and sometimes have long stringy arms, and these are superimposed on an irregular nebulous outline. Their spectra closely resemble those of T Tau stars yet they contain no illuminating star, nor is a suitable star normally seen nearby. Until very recently the origin of their exciting energy was a mystery. They were thought to be regions of star formation and therefore good

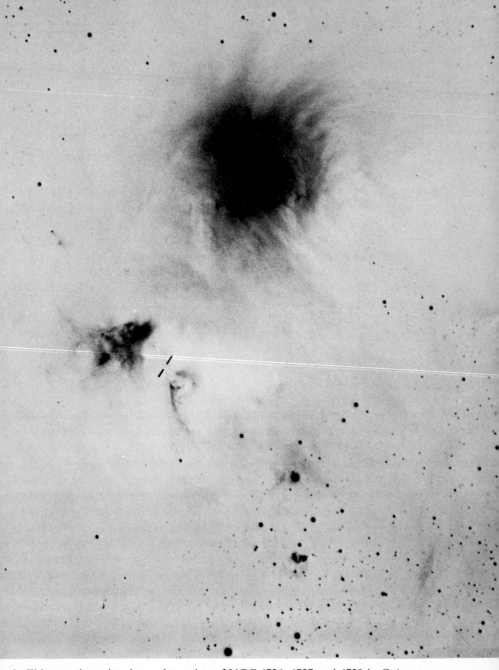

9. This negative print shows the region of NGC 6726, 6727 and 6729 in CrA. An infrared source is marked by the black bars; this source probably illuminates the Herbig-Haro object to its south west.

candidates for infrared sources, but no infrared signal has yet been detected from a genuine Herbig–Haro object. Only if we broaden our definition to include some rather similar objects which do contain faint stars do we find an infrared signal from circumstellar dust clouds. The two principal examples of this latter type of infrared source are Haro 8a (MVP 4 in reference 5) and object number 16 in the Chamaeleon T association.[18]

It is now apparent that infrared astronomers who laboured to find signals from the Herbig–Haro objects were doing their job too well. Had they taken less care in setting their telescopes they might have found the infrared sources, for these lie a little distance away from their associated Herbig–Haro objects.

The first infrared source associated with a genuine Herbig–Haro object was found by the Stroms and Grasdalen in the R CrA group, a small young cluster.[19] This is shown in plate 9, the infrared source being marked by a bar. The infrared sources associated with this and other Herbig–Haro objects seem to be bright stars obscured by intervening dust. The Herbig–Haro objects are reflection nebulae mirroring the T Tau spectra of these stars.

An unusual object which is probably quite young is M1–92. This was discovered by Minkowski on objective-prism plates taken from Mount Wilson during the 1940s; the discovery was announced in 1946. On photographs, and indeed optically through even quite modest telescopes, M1–92 resembles a footprint, as plate 10 shows. The nebula is in two parts – a bright portion about 3" arc in diameter (the 'sole') and a smaller fainter appendage (the 'heel'), separated from the sole by a dark lane. Within the dark lane is a star recorded only in the red and photographic infrared.

The spectrum of the footprint nebula is rather like those of the hotter T Tau stars and of Herbig–Haro objects. Because there is a strong continuum to the spectrum we can say with some confidence that M1–92 is not an emission H II region but a reflection nebula. The sole of M1–92 is a bright object – about magnitude 11. The star it reflects must therefore be very bright, and its absence at visible wavelengths is attributable to extinction caused by dust in the dark lane. With so much dust lying close to the star, M1–92 might be thought a certain infrared source, and indeed it is.[20,12] The colour temperature of the dust is approximately 750 °K, and 92 per cent

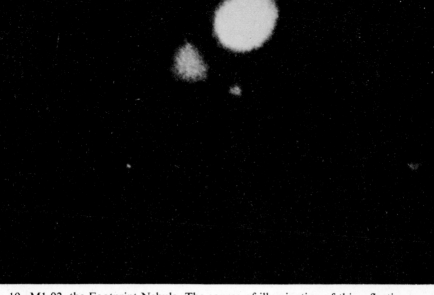

10. M1-92, the Footprint Nebula. The source of illumination of this reflection nebula is hidden in the dark instep.

of the energy of the object is in the infrared component. The extent of the source at infrared wavelengths is not known.

H II REGIONS

Not all H II regions contain young clusters, although all must house at least one hot and presumably young star. They are, nonetheless, regions in which stars may form, and since we expect them to comprise both gas and dust in about the ratio 100:1 by weight, there is every reason for expecting at least some infrared signal from them. And even if the dust is too thinly distributed or too cold to be measured, there is always that old standby, free-free radiation, to fall back on.

At 2 μm most H II regions emit only free-free radiation. The IRC contains several, notably the Messier nebulae 17, 20 and 42. The giant H II region in the Large Magellanic Cloud, 30 Doradus, also

131

has the energy distribution of *Bremsstrahlung*.[21] It is only when we consider longer infrared wavelengths that we find a contribution due to dust, and it now looks as though all H II regions have very pronounced infrared excesses in the far infrared. Some – generally the denser examples – contain dust warm enough to be detected easily at 10 μm, and in a few – the very dense *compact H II regions* – the dust excess is already present at 2·2 μm.

THE FAR INFRARED

Most H II regions, then, can be studied best at wavelengths inaccessible from the ground. The newly exploited 35 and 350 μm windows give some access, and at both these wavelengths the brightest celestial sources are almost without exception H II regions. However, most of our knowledge about the infrared properties of H II regions has come from aeroplane, balloon or rocket observations in the 40–400 μm range.

Low and Aumann in 1970 were the first to publish far infrared measurements of H II regions.[22] They found signals in the 50–300 μm band from M 17 and M 42. They were able to estimate the infrared luminosities these complexes emitted, finding $1·2 \times 10^6$ and $1·5 \times 10^5$ times that of the sun respectively. This is rather greater than the visible and ultraviolet radiation of the known stars: these H II regions therefore emit most of their energy in the infrared. In 1971 Harper and Low added five more to the list of detections, including NGC 2024, the fan-shaped nebula Nf ζ Orionis.[23]

At this time rocket-borne detectors were being used to survey selected portions of the sky. The first of these was by Hoffmann, Frederick and Emery.[24] They found 72 possible sources, and of these at least 50 can be identified with H II regions. Further survey work was undertaken by Emerson, Jennings and Moorwood who favoured the use of balloons to carry their detectors above the atmospheric water vapour.[25] A number of researchers have recently made more detailed observations of some of the brighter sources.[26–28] Combining their results we can deduce the following:

The infrared excess in the extended H II regions is almost certainly produced by dust heated by the early-type stars which are responsible for the ionisation of the region. A typical value for the temperature of the dust is 100 °K; the Orion Nebula is at 70 °K.[26] The dust is almost certainly mixed fairly uniformly with the gas. It must be a relatively stable sort of material with nearly black-body emissivity; graphite would be a fair guess. If, as is usual for small

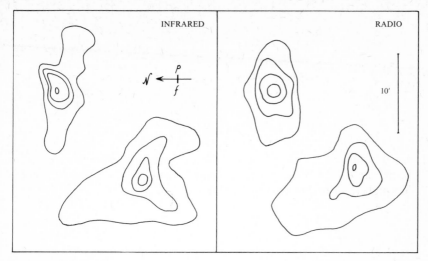

Fig. 28   A comparison of infrared and radio maps of an H II region, NGC 6357. The infrared map, at 100 μm, shows the distribution of the dust. The radio flux arises from the ionised gas.

grains, the emissivity falls off roughly as the inverse square of the wavelength, the amount of dust must be quite large, perhaps in excess of the 1:100 dust to gas ratio considered normal.[27]

The luminosity of the infrared component is closely proportional to the radio flux from the H II region. This is perhaps not surprising: the radio signal is produced by free-free radiation and therefore derives its energy from the ultraviolet radiation of the embedded stars; the dust is powered mostly by the blue and ultraviolet radiation of the same stars. On the other hand an H II region containing very little dust would have a small infrared signal and would not fit the correlation. We must therefore conclude that the H II regions bright enough to have been detected at 100 μm all contain similar amounts of dust.

Some of the larger specimens have been mapped.[25] The infrared isophotes very closely resemble the radio maps (Fig 28), and this tells us that the dust is distributed exactly as is the gas. It must therefore coexist with the ionised gas, something which it can probably do for no more than about a million years. An exceptional case is the Orion Nebula in which the infrared and radio centres are discrepant by about 2′ arc. As we shall see later, the Orion Nebula probably is a rather different object from most H II regions being, in fact, so dense that it is not ionised all the way through.

Some of the 100 μm sources which are certainly identified with radio H II regions have no optical counterparts because they lie behind absorbing clouds of interstellar dust. Others equally bright in the infrared and radio are such prominent optical objects that they were recorded in the NGC or even Messier's list. So we cannot see all H II regions on even the best optical photographs. When we do see the region, frequently its optical isophotes do not well match the infrared and radio maps, and this is because of the irregularities of the intervening dust. By comparing the infrared or radio maps with photographs taken in the Hα line we can deduce the distribution of the intervening dust. At least in some cases much of the dust lies immediately outside the nebula, and some of it is therefore warm enough to contribute to the infrared flux.

THE 10–20 μm REGION

A black body at the sort of temperatures we find dust to assume in H II regions is not too cool to be detected at 20 μm, although its detection there will be more difficult than at longer wavelengths where most of its energy is released. But because of the relative ease of ground-based 20 μm photometry compared to 100 μm work, the first discovery of thermal radiation from an H II region was made at the shorter wavelength. As long ago as 1967 Kleinmann and Low discovered an 'infrared nebula' in M 42.[29] This we now know to be the central part of Orion's infrared cloud; at the time it was the first extended infrared source to be discovered. It is frequently referred to as the Kleinmann-Low nebula, and is often abbreviated to **KL**, a designation which should not be confused with the symbols for the 2·2 and 3·5 μm wavebands. Extended sources have also been found at 20 μm and even 10 μm in M 17.[30]

It transpires that the Kleinmann-Low nebula is not the only extended infrared source in M 42: a second was discovered in 1969. This is centred on θ^1C Orionis, one of the stars of the Trapézium group, and lies about 1' arc north of the **KL** nebula. The Trapezium source is also named after its discoverers, Ney and Allen, and is frequently abbreviated to **NA**.[31] **NA**, then, is the reradiation cloud of the dust that lies near the Trapezium stars, and since it is slightly warmer than **KL** it is an easier object at 10 μm. In fact it is sufficiently bright at 10 μm that Stein and Gillett, who made an independent discovery of the source on the same night but at a different observatory, were able to take narrow-band measurements of it throughout the 8–14 μm window.[32] They found the **NA** source to have a silicate signature.

This remains a unique observation. No other H II region has been found to have silicate emission. Indeed, apart from a muted feature in η Car, no object besides comets and the late-type stars where the grains form has a silicate bump. It is very unlikely that grains would form in an H II region like the Orion Nebula because the density is so low that suitable atoms meet but once in a blue moon. They must therefore have originated around late-type stars which have long since ceased to shine and been gathered together into the material which eventually formed the Orion Nebula. This is evidence for a high degree of mixing of interstellar material.

Messier 8 has also been studied in detail. The dust radiation at 10–20 μm is centred on the star Herschel 36 which lies very close to the 'Hourglass' – the bright central knot of the nebula.[33] There is a weak infrared signal from the Hourglass too, suggesting the presence of a small amount of dust there.

It has now become rather fashionable to make 10- and 20-μm maps of H II regions, and several have recently been published. The normal practice is to choose those with strong radio fluxes, and in particular those in Westerhout's catalogue of extended radio sources. The infrared and radio maps are always quite similar. I will name only W 3 (part of IC 1795), the first of the present band-wagon;[34] this region will feature in a later chapter. Other H II regions will be found described at reference 35.

COMPACT H II REGIONS

If I were asked to define a compact H II region, a term which has grown into the literature in a rather indefinite way, I should frame my answer around just one distinguishing feature: the density. Compact H II regions are quite simply high-density versions of normal H II regions. The higher density manifests itself in their smaller size (frequently only about 1′ arc in diameter), high radio flux for their size, and slightly different optical spectra in which the emission lines which prefer higher densities are enhanced. Compact H II regions are also characterised by their prominent infrared continua which have a higher colour temperature than do the extended examples. Exactly why this should be so, even in higher density conditions, is not entirely clear.

Most of the compact H II regions are known only by their numbers in a catalogue by Sharpless. Only a few are NGC entries. One of these is NGC 7027, the irregular nebula in Cygnus discovered by Webb in 1879 and shown in plate 11. Many books persist in calling this object a planetary nebula, but this is an almost certain

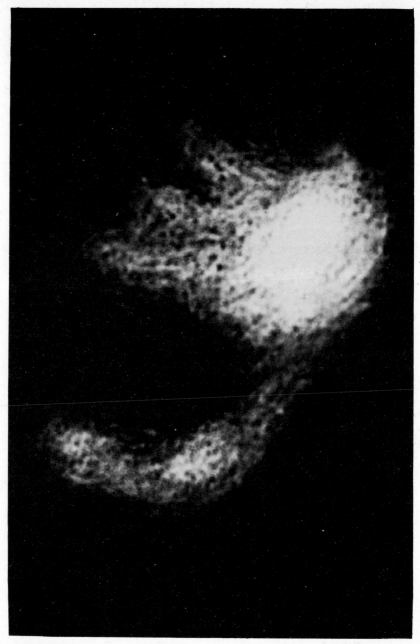

11. NGC 7027 would appear symmetrical but for the dust which partially obscures it and produces the strong infrared emission.

case of misidentification. Other compact H II regions which have found their way into catalogues of planetary nebulae include K3–50 (Kohoutek), H2–3 and H2–6 (Haro), and M1–78 (Minkowski). All four have strong infrared continua.

The excess in NGC 7027 was discovered in 1967.[36] This is a particularly interesting example because the first narrow-band photometry in the 10 μm window showed an emission feature somewhat resembling the silicate bumps to be recognised later in late-type stars. Observations at higher spectral resolution have now been made and it is clear that this feature is not a silicate bump. Carbonates have been suggested instead. NGC 7027 is a prime candidate for infrared spectroscopy and will feature prominently in Chapter 10. Infrared maps of NGC 7027 show a symmetrical structure,[37] as do radio maps. In the visible the outline is irregular. Whether the dust which gives rise to the excess and the visible obscuration is all inside, all outside or both in and out of the nebula is a matter of debate.[37,38]

Historically, K3–50 was the next compact H II region to be studied.[39] Curiously this lies very close to an IRC source, but is not coincident. None of the compact H II regions are bright enough at 2·2 μm to have gained inclusion in the IRC. Energy distributions of K3–50 and NGC 7027 are given in Fig 29.

Most of the compact H II regions examined in the infrared have been studied by Jay Frogel and Eric Persson and published in a succession of little papers.[40] They have found very large infrared excesses just warm enough to be measured at 1·65 μm and rising steeply towards the doped Ge region. The sources are extended, usually being comparable in size with the optical nebulae. The outer portions are much redder than the inner, and K–L indices in excess of 3 magnitudes are common. For stars such large indices are almost unknown. An obvious interpretation of the increase in the near-infrared colour indices further from the centres of the compact H II regions is that the dust there is cooler than near their centres. Which is just what we would expect if the dust is heated by stars at the centre.

One of the objects studied by Frogel and Persson is NGC 6302, sometimes known as the Bug Nebula.[41] This H II region contains a central bright portion which is fairly dense, and a larger and rather unusual area of jets and extensions. The infrared radiation seems to arise mostly from the central region, and the dust radiation is not observed shortwards of 3·5 μm, as Fig 29 shows.

Oddly, an H II region which gives one of the strongest infrared

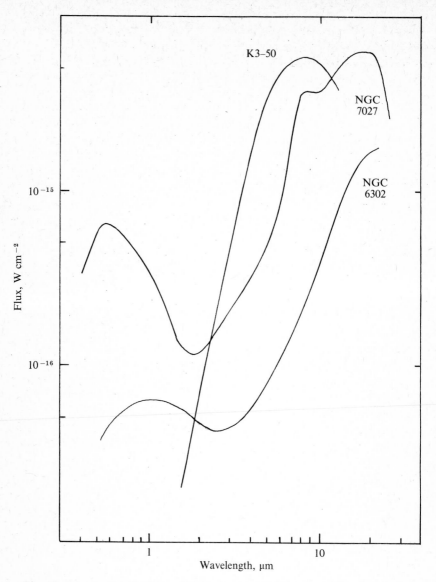

Fig. 29 Energy distributions of three compact H II regions. The dust is about as cool as in most planetary nebulae.

signals at long wavelengths is optically quite invisible. It was first recorded as a radio source and was given the designation G 333·6 —0·2 (from its galactic coordinates). The infrared excess was discovered in 1973.[42] This southern-hemisphere object, which lies in the constellation of Norma, is totally obscured in the visible by the dark clouds of material that throng that portion of the galactic plane. G 333·6 —0·2 is probably the most luminous H II region known and must therefore contain at least one very bright star. It is estimated that the principal star is of spectral type O4, which is about as bright as stars are found on the main sequence.

8

IRTRONS AND ISLAND UNIVERSES

A galaxy is some breccia constructed of all the types of objects we have encountered in the last four chapters. We have found so many of these to radiate unexpectedly strongly in the infrared that we feel justified in expecting galaxies to do so too. But this is a rash prediction, and we should first do a little stocktaking to get things into perspective.

If you set out to write a book about infrared astronomy, you cannot expect much of a readership unless your heroes and heroines have infrared excesses. Nobody wants to know about the stars which have swept their circumstellar dust under somebody else's carpet and now present clean-shaven energy distributions devoid of infrared bumps and wiggles. It is therefore easy to give the impression that the sky is so full of infrared personalities that the only stars with 'normal' behaviour are the ones we use as standards. The truth of the matter is that although infrared excesses are wide ranging in the types of object they favour, numerically they are rare. This is brought home to us very clearly when we examine galaxies and find that their infrared colours are almost exactly what we would predict on the basis of their content of normal stars. Just as the optical spectra of galaxies are in the range of types G to K, because of the great number of late-type stars they contain, so their optical and infrared colours are roughly equivalent to those of G or K stars with the addition of only a few NML Cyg or NML Tau objects.

Probably because of this fact, and also since galaxies are not bright objects, few of them have been measured in the infrared. Only one, Messier 31, is included in the IRC (entry +40 013). It was the indefatigable Johnson who first tackled galaxies at infrared wavelengths.[1] Subsequent research in the PbS region has served only to embroider his findings.[2] And, not surprisingly, M 31 was studied in depth by the CalTech team who made scans across the galaxy to map its infrared radiation.[3] They found the central region to be more prominent relative to the arms in the PbS region than at optical wavelengths,

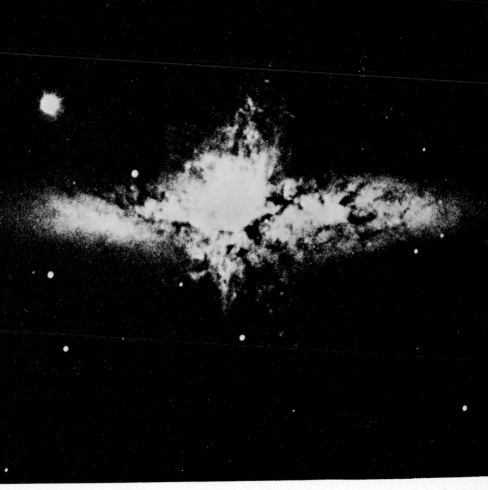

12. At infrared wavelengths Messier 82 is the brightest galaxy.

and this is no surprise since the centre of the galaxy contains mostly late-type (population II) stars while the spiral arms comprise hot young (population I) stars. No hint of an infrared excess was found at L.

The story, however, does not end here. A whole chapter could scarcely be devoted to galaxies and their ilk without there being something more interesting to say about them. Sure enough some galaxies do have infrared excesses. In a very small number this excess is observable in the PbS region. An example of this is the pair of radio-bright galaxies in the constellation of Grus, NGC 7552 and 7582;[4] another is the large, bright edge-on spiral NGC 253.[5] Usually,

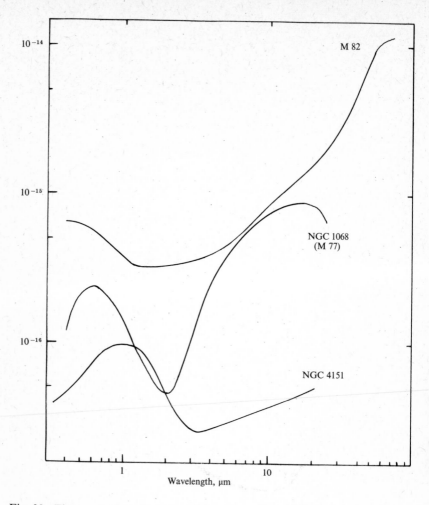

Fig. 30 Three galaxies with strong infrared continua. The dust in Messier 82 is very cool and the energy distribution peaks at about 70 μm. The other two are Seyfert galaxies.

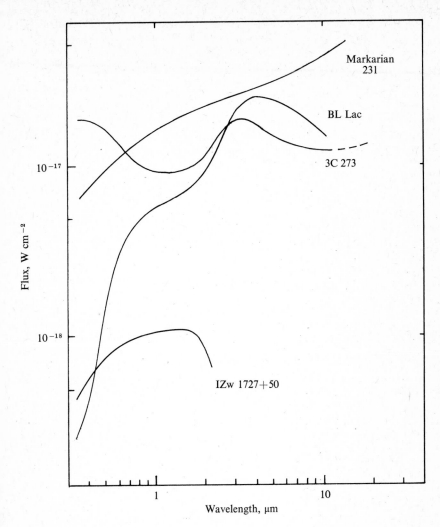

Fig. 31   Energy distributions of some peculiar extragalactic objects. Markarian 231 is the most luminous galaxy known.

however, it is 10 μm photometry which reveals the infrared excess. The brightest galaxies almost attain zero magnitude at 10 μm.

We shall first examine the question: what sort of galaxies have infrared excesses? Only then will the origin of the excess be discussed, although by now the reader may well have a pretty good idea of at least one likely mechanism. It is not a bad approximation to say that a galaxy will probably have an infrared excess if it has any optical peculiarities, either morphological or spectroscopic. Spirals are favoured over ellipticals. A better rule is that if the object is a radio source it will have an infrared excess. The following list will give some indication of the types of peculiarities relevant to infrared astronomers, and examples of the energy distributions of most of these types appear in Figs 30 and 31. It should be noted that the field is still young, however, and there can be no guarantee that the list is in any way comprehensive. Most of the 10 and 20 μm photometry of galaxies has been published in a small collection of papers by Frank Low and his collaborators (see particularly references 6 and 7).

PECULIAR AND IRREGULAR GALAXIES

The galaxies being singled out under this heading do not form a homogeneous group, and it is best to illustrate the type by means of a few examples; some are shown in plates 12, 13 and 14 .The peculiar, seemingly explosive galaxy M 82 is a bright infrared source. In M 51 it is the companion to the south (NGC 5195) which has the infrared excess. Also shown is NGC 5128 (Centaurus A) in which there is an infrared excess originating in a small central core.[8] The strange radio galaxy Cygnus A, once thought to be a pair of colliding galaxies, is also very luminous in the infrared.[7] Finally the peculiar jet from the nucleus of M 87 has probably been detected at 1·55 μm.[9]

COMPACT GALAXIES

Compact galaxies are, as their name implies, small, bright and condensed. Many have recently been found, mostly as a result of the work of the late Fritz Zwicky. On a cursory inspection of photographs they appear stellar; it requires a careful study to distinguish their fuzzy nature. With several thousand to tackle it is surprising that only a very small number have been observed even out to 3·5μm. Indeed only three have been deemed to merit a separate paper. These are 3C 120 and the Zwicky discoveries I Zw 1727 + 50 and IV Zw

13. Messier 51 and its irregular companion NGC 5195.

14. Centaurus A, NGC 5128.

0039 + 40 (the catalogue numbers refer to their coordinates).[10] Two others were included in a review paper in 1971,[11] and several more were contained in Rieke and Low's 1972 survey.[7]

MARKARIAN GALAXIES

Markarian's survey selected galaxies with strong blue continua, which in other words means those with ultraviolet excesses. What he in fact produced was a list of galaxies which are almost guaranteed infrared objects. An energy distribution for Markarian 9 was published in the review paper mentioned above:[11] to 2·2 μm it rises much faster than most other galaxies. Several others have recently been measured out to 10 μm;[7] of these Markarian 231 is the most

146

luminous galaxy currently known, and most of that luminosity is emitted in the infrared.[12]

As long ago as 1964 Johnson made 2·2 μm measurements of 3C 273,[13] and the following year Low and Johnson pushed out to 10 μm finding a substantial infrared excess.[14] The infrared luminosity exceeds that at radio and optical wavelengths by a factor of ten. Claims that 3C 273 varies at 10 μm are probably not yet justifiable: it is a very faint source. No other quasar has been measured at 10 μm, and some cannot therefore have such large excesses as 3C 273. A sample of twenty-eight was investigated at H and K in 1970:[15] all show signs of infrared excesses, and in many the energy distribution rises very quickly into the infrared.

BL LACERTA OBJECTS

It is only in the last few years that a class of extragalactic sources has become recognised of which BL Lac is both the type object and the first discovered. The number in this class is still small, but additions are frequently made. A definition based on the recognised properties would read something like this:

BL Lac objects are radio sources with optical counterparts. They are irregular optical and radio variables, with amplitudes of about one magnitude and characteristic time scales of as little as one day. Indeed BL Lac is not the only one catalogued as a variable star. Their spectra show no emission or absorption lines whatsoever, and no redshift can therefore be determined. Optically they appear almost stellar, except that a small nebulous halo surrounds BL Lac itself, and it now seems that this halo is a galaxy, the very bright nucleus of which is the principal radio and optical source. BL Lac objects have ultraviolet excesses (or so it is said, but what it is in excess of is not clear). And – need it be said? – they have prominent infrared continua. In fact they radiate most of their energy in the infrared.

The first infrared observation of one of these strange creatures was in 1969 when Oke *et al* measured BL Lac itself at H and K.[16] Two years later the prominence of BL Lac at longer wavelengths was discovered,[17] and thereafter observations of other BL Lac objects came rapidly.[18]

147

It would not be an overstatement to say that of all extragalactic objects, Seyfert galaxies are the most revered by the infrared community. When Neugebauer *et al* wrote their 1971 review, 60 per cent of the extragalactic objects detected at 10 μm were Seyferts, and they still feature prominently in all lists of infrared detections.

Seyfert galaxies were classified long ago on the basis of their compact bright nuclei and their rich emission-line spectra. Seyferts have broad emission lines, but the class can probably be extended to include otherwise similar galaxies with narrower lines. Emission lines are, of course, something one associates with infrared excesses in galactic objects. So it is gratifying to find that the same correlation seems to exist in the case of galaxies even though it is difficult to understand the conditions under which the emission lines are formed and to be certain of the cause of the infrared excess. Seyfert galaxies also have ultraviolet excesses.

It was Pacholczyk and Wisniewski who started the infrared Seyfert bandwagon in 1967,[19] and this was quickly followed up by Low and Kleinmann.[20] The latter authors soon determined that the brightest Seyfert galaxy at 10 μm is NGC 1068 (which for some odd reason is never known by its Messier number, 77), closely followed by NGC 4151. A lot of effort has been put into investigating these two more thoroughly with particular reference to the angular size of the infrared emitting region and to the possibility of variations at infrared wavelengths.

NGC 1068

In 1971, Neugebauer *et al* studied the size of NGC 1068 at H, K and N.[21] They found the galaxy to be extended at the shorter wavelengths, probably due to the contribution of starlight; but at 10 μm they were unable to resolve it. By 1973 Becklin *et al* had so improved their resolution, by using small focal-plane apertures on the 200-inch telescope, that they succeeded in resolving NGC 1068 at 10 μm, finding it to be about 1″ arc across.[22] In physical terms this means that essentially all the infrared radiation originates within a sphere of diameter 60 parsecs; this is pretty small when one considers that galaxies are usually several kiloparsecs across. For comparison, M 82 at 10 μm measures 25″ × 8″ arc, yet it too has something of a core.[23]

The first claim that NGC 1068 varies in the infrared, at K, was made in 1970, but this is generally discounted on the grounds that the author (who I will not name) mis-centred the galaxy in his beam on a couple of occasions and thereby measured a spuriously low flux. More recent papers disagree about the reality of 10 μm variations, Low and Rieke believing in them,[24,25] Stein, Gillett and Merrill rating them below the statistical accuracy of the data.[26] On the evidence so far presented, this author favours the latter view.

NGC 4151

The consensus of opinion is that NGC 4151 does vary in the infrared. At the short wavelengths the variations almost certainly are real.[27,28] At 10 μm the safest verdict is that there is enough evidence of variation to warrant further investigation.[25,26,29] There has been no attempt at resolving the infrared component of NGC 4151, but Penston *et al* made an interesting deconvolution of its energy distribution into two components, one a normal galaxy, the other a compact nucleus emitting most of the infrared radiation.[28] This nucleus is, apparently, less than 5″ arc diameter.

ORIGIN OF THE INFRARED EXCESS

Before we can discuss the nature of the infrared emission in galaxies, we must examine the optical and radio radiation they emit. The radiation from the stars they contain clearly accounts for the visible and infrared radiation in the normal galaxies with which this chapter opened, but it cannot explain the infrared excesses in Seyferts and their peers. Nor can starlight alone produce any detectable radio emission. Hot stars surrounded by gas will excite free-free radiation which can have a radio component if the gas is not too dense. However, this mechanism can in most cases be ruled out because free-free is far too inefficient a mechanism to produce the sort of radio signals found from optically faint galaxies. More important, the energy distribution observed in radio galaxies and quasars is totally inconsistent with that produced by free-free. Whereas free-free radiation falls with increasing wavelength, in many radio galaxies the flux initially increases with wavelength. In this respect the radio observations follow a similar pattern to the infrared, and it is tempting to attribute the same mechanism to both. This temptation is strengthened by the apparent existence of a tight correlation between infrared and radio luminosity.

It is now clear that radio galaxies and quasars give off synchrotron radiation, those piercing shrieks emitted by high-energy electrons when they embark on their helter-skelter paths in magnetic fields. To maintain synchrotron radiation one requires a continuing supply of high energy electrons; where these come from is by no means clear.

Synchrotron radiation extends right down into the optical in many cases, thus we are well on the way to explaining infrared excesses, or so it seems. Unfortunately, if we extrapolate the radio spectrum back into the infrared it generally passes too low. The component which produces the radio emission cannot account for the infrared too, and we must invoke a second, denser, nuclear component which gives rise to the infrared radiation but which turns down at longer wavelengths, rather like free-free does, so as not to contribute in the radio. It was once thought that this turndown had to be extremely steep, for Low and Aumann flying in aircraft made apparently reliable detections of NGC 1068 and M 82 at 100 μm.[30] To drop from the the high peak at 100 μm to the radio continuum at a few centimetres wavelength stretches the synchrotron interpretation to its logical limit. Some might say that it was stretched beyond its logical limit when in 1970 Low proposed the existence of *irtrons*, clusters of tiny nodules of synchrotron radiation in the nuclei of galaxies.[31] Low got over the problem of their energy source by assuming that his irtrons were the site of creation of matter in a conveniently modified Steady State universe.

Irtrons would not be so necessary if only the giant 100 μm fluxes were not real. So it was with some relief to the astronomical community that the 100 μm observations of NGC 1068 were finally retracted.[32] The original 'detections' at 100 μm had been of part of the aeroplane! Unfortunately this same paper reported 100 μm detections of M 82 (at a much lower flux) and NGC 253, so the problem still survives to a lesser degree. However, irtrons as such are not much talked about now, even if synchrotron radiation finds a number of supporters.

In the case of NGC 1068 the synchrotron model is in particular trouble. First, it is impossible to maintain strong magnetic fields over the whole 60 pc. core. If individual irtrons are involved they must be distributed so as to look like an extended source. Secondly, recent ground-based observations beyond 20 μm show, not a continuing rise, but a steep drop not consistent with synchrotron radiation.[33]

Thus we may have to fall back on that hoary chestnut, dust. There is certainly no difficulty in distributing the dust in a 60 pc. core. However we may be faced with another problem, namely the variability of the sources. Dust, after all, does not radiate by itself – it merely transforms incident high-energy radiation to lower energy. Somewhere there must be a source of energy powering the dust reradiation, and it is this source which must vary. A change in the radiation of the central object moves outwards at the speed of light reaching all the dust grains of a given temperature at the same time. The temperature of these grains therefore changes, causing a variation in the infrared flux. However, the new infrared flux from the dust grains must now travel to the observer on earth, and the radiation from the near side of the dust cloud will arrive before that from the far side. The result of this is to smooth out the change from an instantaneous event to one with a characteristic period roughly equal to the light travel time across the dust cloud. Now, we see that if the typical time scale of variations is of the order of a day, as has been suggested for at least the BL Lac objects, the infrared source must be not much bigger than the solar system. Patently we cannot expect to find any normal object surrounded by a dust cloud of that size but radiating 10^{11} or more times as much as the sun. And if we make it, instead, a million very bright stars each in its own dust cocoon we face the statistical argument that we should not expect sufficient of them to vary in phase for the nett signal to vary at all. This is also an argument against the existence of numerous small irtrons.

One of the mechanisms occasionally mooted for providing $10^{11}L\odot$ in a small volume is the gravitational collapse of the galactic nucleus to form a black hole. If this were the case, the dust itself would be spiralling into the gravitational well and radiating in the infrared the energy it released on its journey. But such a model is no more than speculation, and the truth of the matter is that we really don't know whether the infrared excesses in galaxies are caused by dust, synchrotron radiation or some more esoteric mechanism. This is one of the great outstanding problems in infrared astronomy.

THE MAFFEI OBJECTS

In 1969 the Italian astronomer Paolo Maffei discovered two faint nebulous objects in the Milky Way in Cassiopeia on a plate he had taken in the photographic infrared.[34] These soon became known as Maffei 1 and 2. Both are invisible on blue photographs, and it was thought that they might be small H II regions reddened by intervening dust. It therefore came as some surprise when a thorough study of the two, published in separate papers[35,36] showed them to be galaxies. Lying as they do almost exactly in the plane of our galaxy, they experience heavy reddening so that their very existence is unsuspected at optical wavelengths. Yet the brighter of the two, Maffei 1, would be almost as prominent as M 31 if the intervening dust were removed. Maffei 1 and Maffei 2 are two of the most massive and luminous members of our Local Group of galaxies.

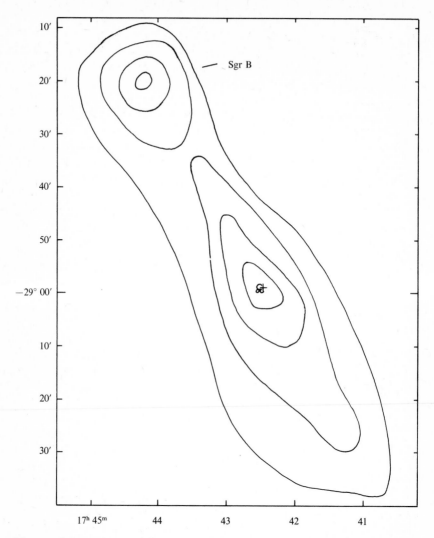

Fig. 32 Schematic map of the centre of our galaxy. The contours show the 100 μm emission and include the H II region and radio source Sgr B which is not recorded at shorter infrared wavelengths. Three sources cluster at the centre and are marked by circles. The largest circle represents the most luminous source which lies at the dynamical centre of the galaxy. The cross is a point source – probably a distant late-type star.

Infrared excesses are things we tend to regard as status symbols. There is almost a cult built around them: the infrared astronomer feels a warmth towards objects with particularly big or unusually cool excesses. Indeed he should, for it is these which keep him in business. The nuclei of peculiar galaxies are examples of such objects, and there is surely no infrared astronomer alive who does not speak with affection of NGC 1068 or 4151.

Our own galaxy, as best we can tell, is not peculiar in any of the right ways. Much as it would pander to our anthropocentric pride if it too had an infrared excess, we have no justifiable reason for expecting it to have. But the centre of our galaxy is appreciably nearer than that of any other island universe, and it is therefore an obvious object to examine at infrared wavelengths. Becklin and Neugebauer thought so too: in 1968 they published observations out to L,[37] and in 1969 to 20 μm.[38] A 10 μm map of the galactic centre was made by Rieke and Low in 1971.[39] Fig 32 combines these data into one infrared map, and it will be seen that the situation is by no means simple. We can distinguish several components.

The stellar population

In the 2 μm region there is an extended area of emission about 1° across which peaks quite sharply towards its centre. This is similar in shape and profile to the nucleus of M 31 at the same wavelength, and is due to the concentration of late-type stars at the centre of our galaxy. It coincides closely in position with the deduced dynamical centre of the galaxy and also with the synchrotron radio source Sgr A which is generally identified as the true centre. However, the infrared source does not have the colours of normal stars, being much redder. Indeed it is apparent that this must be so for there is no optical counterpart to the infrared source. Becklin and Neugebauer identified the cause of the discrepancy as the ever-present inter-stellar dust.[37] Over the 10 kpc path to the centre of the galaxy there is sufficient dust to extinguish the visible radiation by no less than 27 magnitudes, or about a factor of 10^{11}. But 27 magnitudes extinction at V produces only a couple of magnitudes dimming at K, so the radiation is easily detected in the infrared.

A nearby point source

A bright point source lies about 10″ arc away from the centre of the extended infrared source. It has similar infrared colours to the extended source, and the most natural explanation is that it is a particularly bright late-type star near the centre of the galaxy also experiencing 27 magnitudes of visible extinction. A star such as α Ori would be sufficiently bright. The source could, however, be a foreground NML Cygni object.

Sgr B

Radio maps of the galactic centre show not merely a single synchrotron source (Sgr A) but also two satellite sources, one on each side along the galactic plane. The source to the SW of Sgr A is rather weak and not always considered to be separate. That to the NE is quite clearly distinct and is known as Sgr B. This is not a synchrotron but a free-free source, and it is probably a giant H II region near the galactic centre. One section of Sgr B is a rich neutral cloud of molecules: a great variety of molecular emission lines have been found here by microwave observers. Molecular clouds will feature in the next chapter; for the present it is probably the H II region which concerns us. Like the major H II regions described in Chapter 7, Sgr B has pronounced infrared emission, but in this case the emission has been detected only at 100 μm,[40] and is not visible at 10 μm.[30]

The core source

Becklin and Neugebauer found a particularly red source at the very centre of the extended 2 μm region. This object is not stellar, but about 15" arc in diameter. Shortward of 2 μm it cannot be isolated from the extended background, but beyond 5 μm it is by far the brightest component of the galactic centre. From 10 to 20 μm the energy distribution of the core source rises more steeply than that of most other galaxies, even Seyferts. Our galaxy *does* have an infrared excess.

At 10 μm Rieke and Low were able to resolve the core source into three components. Beyond 20 μm, however, observations have not been made with such high resolution and we do not know whether the core source contributes all the flux. At 100 μm there is plenty of flux. The centre of the galaxy is brighter even than Sgr B.[40] The announcement of the detection of the galactic centre at 100 μm, made by Bill Hoffmann, was one of the highlights of the December 1968 meeting of the American Astronomical Society at Austin, Texas; Hoffmann's ten-minute paper was frequently interrupted by the clicking and whirring of cameras. Fig 33 illustrates the energy distribution of the core source and Sgr A in the radio and infrared regions of the spectrum, and it will be seen that the peak of

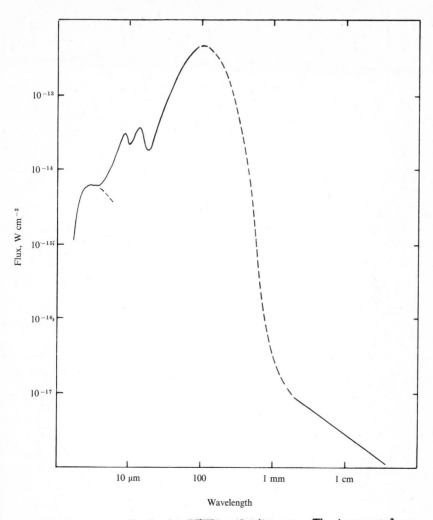

Fig. 33 The energy distribution of the galactic centre. The bump at 3 μm (extreme left) is the stellar component. The linear section at right is the radio synchrotron source. All the remainder is the infrared excess of the core source. The dips at 10 and 20 μm are due to silicate absorption.

the infrared emission is several orders of magnitude higher than the radio and near infrared continua. Rocket observations at 5, 13, 20 and 100 μm show that the Becklin-Neugebauer 15″ arc source is only the central portion of an extended region of emission 500 pc in diameter.[41] The infrared core of our galaxy is therefore larger than those of NGC 1068 and, probably, many other galaxies.

It is not clear why our galaxy should have so prominent an infrared excess, for it is not obviously peculiar. When we total the luminosity, however, we find it to be about 3×10^8 L$_\odot$, and this is a couple of orders of magnitude weaker than the typical Seyfert galaxy. If our galaxy were placed at the distance of NGC 1068 we would not have detected its infrared excess. Thus our suspicion is roused that the infrared phenomenon is a feature of most galaxies – or at any rate most spiral galaxies – and is merely more pronounced in Seyferts, Markarian galaxies, peculiar galaxies and quasars. It is worth adding at this point that no infrared emission has been found from the nucleus of either of the Magellanic Clouds.

Soifer and Houck,[42] in reviewing the nature of the infrared excess in our galaxy, were able to eliminate synchrotron radiation because of the high magnetic fields needed. They likened the 500 pc core to a giant H II region ionised by the hotter stars of the galactic nucleus, and they attributed the infrared radiation to dust grains within this H II region. For our galaxy this seems plausible. Whether Seyferts and quasars contain enough hot stars to produce their infrared excesses by the same mechanism is uncertain.

100 μm SOURCES

If most galaxies may have strong 100 μm emission, they should obviously be examined at this wavelength. However, the difficulties of observing in the far-infrared have so far prohibited long-integration runs on most galaxies. The sky has not yet been covered by 100 μm searches: most have been directed at the galactic plane where H II regions are easily found. A few unidentified 100 μm sources (which may or may not be spurious) have been discovered outside the galactic plane,[43] and further research may eventually reveal these to be galaxies.

THE FAR-INFRARED BACKGROUND

At millimetre wavelengths there is a weak uniform signal from the sky which originates outside our galaxy. The brightness temperature of the radiation is 2·7 °K, and although its cause is uncertain, one theory is that it is the residual radiation from the 'Big Bang' with which our universe may have begun. Whatever the origin, it is important to determine whether it is optically thick – ie whether the brightness and colour temperatures are the same. To do so we must measure it on the short wavelength side of its black-body peak, in the far infrared.

It is, of course, necessary to make such a measurement from above the earth's atmosphere for this both absorbs incoming radiation and emits in its own right. All observations have been made from rockets at 150–200 km altitude. The first such measurement dates from 1968 and was confirmed in 1969.[44] Despite a subsequent downward revision of the absolute calibration,[45] the measured flux in the 400–1,300 μm band (0·4–1·3 mm) is higher by a factor of 40 than that expected from a 2·7 °K black body. The far infrared and millimetre observations cannot be reconciled with any black body.

A more thorough measurement was published by Pipher *et al* in 1971.[46] Using an 18 cm f/0·9 prime-focus telescope these researchers measured the flux from 2 square degrees of sky in the wavebands 70–130 μm, 200–450 μm and 400–1,500 μm. In the two shorter windows they obtained only upper limits, both consistent with a 2·7 °K black body. At the longer wavelength they confirmed the high intensity found earlier, although revising the measured flux downwards by a further factor of two. They were able to eliminate scattered earthlight and atmospheric emission as the causes of the signal, and found no significant difference in the intensity in and out of the galactic plane.

It is now apparent that this far-infrared background is not that from a black-body type continuum. Rather it must be originating in a strong emission line somewhere around 900 μm.[46,47] But whether the emission line originates within the solar system, the galaxy or the intergalactic medium has yet to be determined.

9

HIDDEN SOURCES

So far we have concerned ourselves with visible celestial objects: we have found all of them to emit infrared radiation, and many to radiate at surprisingly high levels. But this is only half the story. We should also admit the possibility that some objects emit in the infra-red but not at optical wavelengths. The IRC contains a few such objects – CIT 1 and 10 for example – and this fact alone virtually guarantees that there be other fainter examples. There are sources too which, while bright at longer wavelengths, are undetectable at 2 μm. And there are sources which are bright in the infrared but which at optical wavelengths, while visible, are so faint that they would forever escape our attention if we never strayed from the narrow visible waveband. Until we have explored these aspects we cannot claim to know the infrared sky.

The IRC provided us with a sample of the brightest 2 μm sources north of $-33°$ declination. Most of these have now been examined in greater detail and we have a fairly clear picture of the content of the 2 μm survey. The great majority of IRC sources are late-type stars, many of which have girdled themselves in draperies of dust. Normal bright stars make up the bulk of the remainder, and only a minute percentage remains to be accounted for by the oddities of the infrared sky. As we have seen in the last three chapters, these oddities include a few H II regions and globular clusters, one galaxy, one Wolf–Rayet star and one or two young objects. The brightest planetary nebula falls faint of the IRC by over one magnitude, and several forbidden-line stars failed to gain inclusion by a smaller margin. Further, most of the oddities have K magnitudes below $+2$, so that they but narrowly gained the distinction of an IRC number. These statistics lead us to suspect there to be many more interesting or unusual objects at slightly fainter K magnitudes, sources which would be found by a deeper survey. To make a survey much deeper than the IRC, however, involves using a smaller beam, which in turn requires either the construction of or use of a different telescope from the 62-inch survey instrument. Since no better infrared telescope has been or is likely to be built, the prospective observer must make use of existing optical telescopes. Even if all the optical telescopes in the world were commissioned for the sole purpose of making a deep 2 μm survey, it would be several years before the task could be completed. Instead, therefore, we must be selective and examine only those regions where we might expect our chances of success to be highest. We must, in fact, bias our sampling.

But a 2 μm survey, however deep, would not find for us infrared sources of colour temperature a few hundred degrees Kelvin. In just the same way no survey at V could possibly have anticipated the IRC. The United States Air Force survey at 4, 11 and 20 μm certainly goes a long way towards finding the cooler sources. But although it has recorded many new objects, and although it will be several years before all of these are examined and understood, the USAF survey contains no more entries than there are 2 μm sources brighter than $+2$ at K. Seen in this light it can have done no more than scratch at the surface. There is plenty yet to find. We are, in essence, barely emerging from the Herschel era of infrared astronomy.

At the time of writing, a small portion of the sky has been explored in a search for infrared sources. H II regions and dark nebulae in our galaxy comprise the bulk of the areas selected for such searches, and new sources are continually appearing in the literature. This chapter can do no more than present a snapshot of a constantly evolving scene; by the time the book appears in print the discoveries and the interpretation of the sources so found may have far outstripped the current state of affairs. With this in mind the reader is asked to refer to the second appendix at the end of the book where important new discoveries which become known in time will be detailed immediately before the book goes to press.

SOURCES WITH OPTICAL COUNTERPARTS

It generally takes two people to make a successful search. One works at the guiding eyepiece and drives the telescope in such a way that the detector sweeps out a raster pattern across the chosen region with each scan separated by rather less than the diameter of the beam. The other watches the pen recorder and calls out whenever a deflection of the pen signifies that a source has been driven into one or other beam. The search is made with as large a beam as prevailing sky conditions will permit, but once a source is found it is centred in progressively smaller beams. Finally, when the observer is satisfied that he has the source right in the beam, he pushes in the mirror allowing him to view the field through the acquisition eyepiece. This is the first test: is there a visible star in the beam?

More often than not there will be a star. This is not necessarily a disappointment, for the star may have a dust cloud which causes its strong infrared emission. The second test, then, is to measure it at two wavelengths, say H and K. If the H-K colour is small, like those of the standards one uses, then the star is almost certainly a common-or-garden red star, or, more rarely, a reddened star, and is probably of little interest.

If no star is visible through the telescope, say to a limit of 15th or 16th magnitude, there is a chance that the source is more unusual. Once again the H-K colour will indicate whether it is merely a particularly red star (H-$K < 0^{m}\cdot5$) or a true infrared source (H-K typically $> 1^{m}\cdot0$). Only in the latter case does the observer really feel he has discovered something worthwhile A great feeling of elation accompanies the discovery of such a source, a feeling which correlates strongly with the H-K (or other) colour index: this is astronomy at its very best.

Subsequent examination of optical photographs may reveal that there is after all an optical object associated with the infrared source. Usually, if H-K exceeds 1 magnitude, there will not be. One particularly interesting pair of infrared sources was found by Allen in a small H II region in Auriga known variously as M1-82 and Sharpless 235.[1] The two sources are identified in plate 15 which shows the region in the light of Ha.[2] Source number 1 lies in the nebulous patch and has no associated star. Its luminosity must be at least as great as that of the brightest visible star in the H II region, seen at the top right of the plate. If it is a single object, it must have the luminosity of an O star.

The second source is a magnitude fainter and lies somewhere in the tiny cluster of objects marked. There may, indeed, be several contributing sources not resolved by the present photometry. A number of the stars in this little group are very red at optical wavelengths. Additionally, the group contains a curved arc of reflection nebulosity (lower left of its brightest member) which must be illuminated by one of the stars. This is rather similar to the nebula attached to Z CMa.

Both the infrared sources have H-K colours of about 2·2 magnitudes, corresponding to colour temperatures of about 850°K. The infrared radiation is almost certainly caused by dust clouds surrounding the stars. The luminosities involved, the degree of infrared

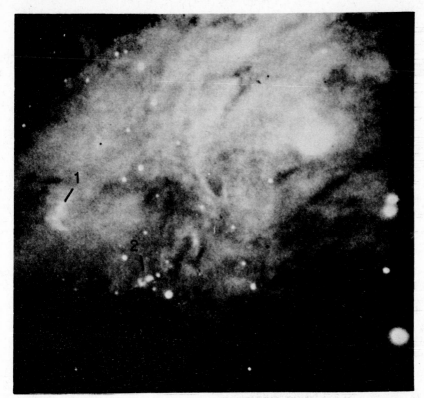

15. The infrared sources in the H II region M1-82.

excess, the position in an H II region, and the optical appearance
and spectra of the objects lead us to believe that they are very young
stars which have only recently formed. They may be younger than
T Tauri stars or the Ae and Be stars associated with nebulosity, and
they may even be the youngest objects yet found. Had the region
not been searched in the infrared, however, they might never have
been noticed, remaining anonymous fuzzy blobs on deep photographs.

INFRARED SOURCES WITHOUT OPTICAL COUNTERPARTS
These are undoubtedly more interesting if only because they remain
entirely the domain of the infrared astronomer. But they are by the
same dint more enigmatic and perplexing, for the lack of optical
data renders their understanding more difficult.

In many cases we can say that the absence of an optical counterpart
is because of extinction by dust. The extinction may be circumstellar,

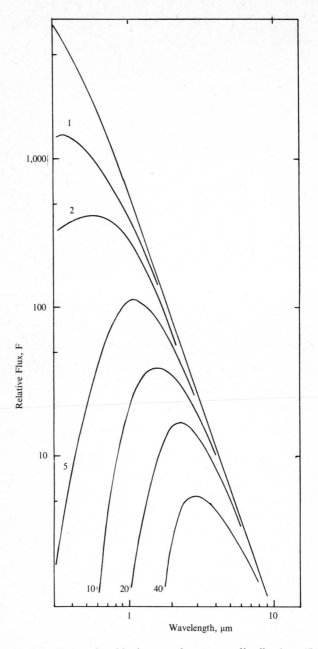

Fig. 34   The effects of reddening on the energy distribution of a hot stellar continuum. Numbers accompanying the curves are values of the visual extinction, A_v, in magnitudes.

in which case the same dust produces the infrared emission; or it may be interstellar. In the latter case there is usually evidence of an intervening dark cloud which on deep photographs manifests itself by the striking absence of faint stars. The Great Rift in Cygnus is such a cloud: behind it are a large number of stars which, in the absence of the dust, would be among the brightest in the constellation and are therefore prominent in the infrared where the extinction is negligible. The largest-known extinction in this region is 12 magnitudes at V for the star known as Cygnus OB2 no 12 (IRC+40 430); this, however, is a visible star. In other parts of the sky the extinction may be higher, and one such region is the Ophiuchus dark cloud in which Grasdalen and the Stroms found no less than forty-one $2 \cdot 2 \mu m$ sources all without optical counterparts.[3] These sources have small $H\text{-}K$ colours and are probably reddened stars without circumstellar dust clouds. The radiation absorbed by the intervening dust must come out at some wavelength: recently a strong $350 \mu m$ source has been found here.[4] This is probably caused by radiation from cool dust in the dark cloud at an appreciable distance from the embedded stars.

Another source in which reddening must play some part lies near IC 2087, a small reflection nebula in the Taurus dark clouds.[1] The star which illuminates IC 2087 is unseen on photographs and is almost certainly the infrared source. It has a colour temperature of about $1,100°K$, and while this could be produced by a circumstellar dust cloud, it is more reasonable to suppose that the intervening dark cloud causes most of the redness of the object. If this is so, A_V for the source could be as high as 30 magnitudes. There must, of course, be little or no obscuration between the star and the nebula it illuminates.

To optical astronomers an A_V of 30 is unthinkable simply because no source experiencing such a figure would be visible. The value for Cyg OB 2 no 12 was long thought to be about as large as would ever be detected. The advent of infrared astronomy has forced a reappraisal of the situation, and astronomers now speak quite freely of visual extinctions of 50 or even 100 magnitudes. Fig 34 illustrates the effect of large interstellar or circumstellar extinction on normal stars, and it will be seen that for A_V in excess of 30 or 40 magnitudes the star cannot be detected even in the photographic infrared.

The energy distribution of a reddened star does not differ greatly from that of a black body. Thus there are two interpretations possible for any infrared source without an optical counterpart. In one the extreme redness of the object is attributed to extinction by inter-

163

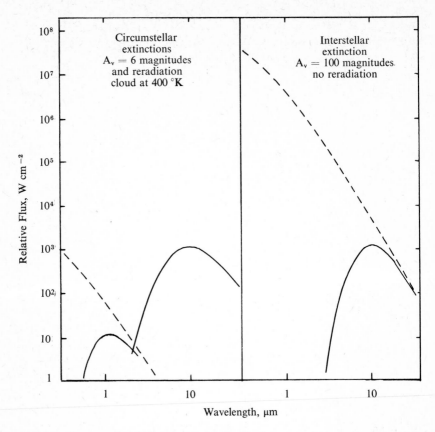

Fig. 35   Very similar infrared energy distributions can be produced by dust emission, as at left, or by extreme reddening of a very luminous star, as at right. In the former the relatively small circumstellar extinction can lower the visible continuum below the level of detection.

vening dust: the infrared photons originate in the photosphere of a normal star and if the dust were removed we would see that star. In the other the object is intrinsically very red and cool, and the infrared radiation originates in a cloud of hot dust. A star may be at the centre of the dust, in which case the dust also serves to extinguish it. But the extinction need be only high enough to render the star optically invisible. An A_v of 15–20 magnitudes would be ample. Fig 35 illustrates how a source with essentially similar colours may be generated by either mechanism.

    In the second alternative the underlying energy source may be a

16. Messier 42 at 10 microns. The visible nebula is centred on the lowest portion which includes the Becklin-Neugebauer, Kleinmann-Low and Ney-Allen sources. The Orion Molecular Cloud 2 is at the top and Messier 43 (NU Ori) is the central source.

protostar, by which we mean that it has not begun to burn its hydrogen but is in the stage of collapsing and warming up. Protostars are the scarlet pimpernels of the stellar world: everybody feels they must exist, but nobody knows how to find or recognise them. Since protostars would be expected to radiate entirely in the infrared, it is popular to suppose that infrared sources without optical counterparts are examples of just such extreme stellar youth.

And for the remainder of this chapter I must steer a course over controversial ground.

THE ORION COMPLEX

The controversy centres on the Orion Nebula. In Chapter 7 I made only passing reference to the infrared emission from M 42; it is actually the greatest complex of infrared sources yet discovered. Many of the visible stars have circumstellar dust emission, mostly at 10 and 20μm. One of the brightest reradiation clouds is the so-called Ney-Allen source (**NA**) which surrounds the Trapezium cluster on the star θ^1C Ori.[5] But long before these sources were known, in 1967, an even more interesting object was discovered by Eric Becklin, then a research student at CalTech.[6] This source, variously known as Becklin's object, Becklin's star, the Becklin-Neugebauer source and **BN**, lies in the bright nebulosity roughly 1' arc south of the Trapezium. It has a colour temperature of about 600°K, and as far as can be ascertained in so bright a nebula, there is no stellar counterpart in the optical or photographic infrared.

Becklin and Neugebauer measured this source out to 10μm where its magnitude is approximately -2. Even in the early days of infrared this promised to be easily measured at 20μm, and such a measurement was attempted by Kleinmann and Low. They found the job impossible, not because Becklin's object was too faint, but because it was swamped by that very bright extended source the Kleinmann-Low Nebula (**KL**; Chapter 7).[7] This was then the coolest-known object, having a colour temperature of only 70°K, and at the time it could not be detected at 10μm.

It is only quite recently that high-resolution maps have been made of the **BN-KL** region; not only have both sources been measured at 10 and 20μm, but a number of fainter and redder point sources have been resolved in the complex.[8,9] Plate 16 has been constructed by combining the maps of references 5, 8, 9 and 10 to this Chapter, together with data from reference 11, and it represents the Orion Nebula at a wavelength of 10μm. There are two basic regions of infrared emission; the northerly was discovered in December 1973 and is known as OMC 2 – the Orion molecular cloud no 2.[10] Some of the sources, including Becklin's object, are unresolved with a $2''$ arc beam.

In order to understand the Orion complex we must combine data from all available parts of the spectrum. First, in the optical, we see a large ionised hydrogen cloud centred on the Trapezium and surrounded by a rich cluster of T Tauri-like young stars. We can tell that there is a lot of dust in the cloud because we see its patchy obscuration causing the whole nebula to be mottled. In places we see quite striking obscuration, as in the dark 'Fish's Mouth' near the Trapezium where a fold of dark material curves part way round the front of the nebula, and further afield but in the same complex, the Horsehead Nebula.

In the infrared we again find evidence of dust in the form of excess radiation from many of the stars. We also find a cluster of infrared sources immersed in an extended region of 20μm emission. At longer wavelengths the Orion Nebula radiates strongly, as do all H II regions, but this radiation is centred not on the Trapezium stars or the radio H II region, but on the **KL** nebula. The radiation peaks at about 50μm, has a total luminosity of about a million times that of the sun, and is clearly attributable to dust within the nebula. The energy radiated by the dust is certainly as high as the Trapezium and other stars can contribute, and is probably much higher, suggesting the presence of other stars unseen due to intervening dust.

At radio wavelengths there is a continuum emission due to free-free radiation from the H II region. In the microwave, where certain molecules emit, there are also strong signals. Several tiny sources of OH emission lie in the cluster of infrared objects, and their significance is not fully understood. In the spectral lines of such molecules as carbon monoxide and formaldehyde there is strong emission from an extended region again centred on the **KL** nebula. From the intensity of these lines we can determine the conditions deep within the nebula and, presumably, where the infrared dust emission originates. We find two surprising things: first, the hydrogen is neutral; secondly, the cloud is very dense. Knowing the density and extent of the molecular cloud we can make an estimate of the amount of dust it contains, and from this estimate we can deduce the extinction on a source lying directly behind the Orion Nebula to be in excess of 100 magnitudes at V.

The picture of M 42 that now emerges is a very dense and perhaps roughly spherical cloud of gas and dust. Optically this cloud would be quite undetectable except for its extinction, but in the case of the Orion Nebula it so happens that there are some bright stars, the Trapezium cluster, which formed near the outer edge on our side of the nebula and have burnt away most of the dust in front of them. These stars ionise the layers we see, making them visible. Whether there are bright stars in the middle or at the far side we cannot tell by optical or radio means.

How do Becklin's object and the neighbouring sources fit into this scheme? This is where the subject becomes particularly controversial. For the first few years after its discovery the **BN** source was assumed to be at its colour temperature of 600°K, and it was frequently supposed to be a protostar. When the molecular observations were published it became apparent that an interpretation based on reddening was also a possibility, and such a model was proposed by Penston, Allen and Hyland.[12] They postulated that Becklin's object is a normal star of very high luminosity which has recently formed in the middle of the molecular cloud and which experiences an A_v of 70–80 magnitudes. They further argued that the dust which absorbs the visible radiation is so distant and cool that it does not reradiate significantly at wavelengths less than about $15\mu m$, and they identified the reradiation cloud with the Kleinmann-Low nebula and the extended $50–100\mu m$ source.

At the time the heavy reddening model for Becklin's object appeared very attractive and it was widely accepted. No model is really worth its salt unless it makes a prediction, and the heavy reddening model did just that. With the knowledge that silicate material is present in the Orion Nebula, at least in the Ney-Allen source where it appears in emission, Penston *et al* noted that if Becklin's object is heavily reddened it should exhibit a silicate feature *in absorption*. If, instead, Becklin's object emits by the thermal radiation of 600°K dust, the silicate feature, if present at all, should be in emission.

The challenge was taken up by Gillett and Forrest who obtained a spectrum over the available portions of the range 2·8 to $13·5\mu m$.[13] Their energy distribution is shown in Fig 36 and it will be seen that the smooth continuum is interrupted by two absorption bands each about one magnitude deep. One of these, at $10\mu m$, is the silicate feature predicted by Penston *et al*; the other, at $3·2\mu m$, is due to ice.

Further evidence favouring a large A_v for Becklin's object came from polarisation measurements between 2 and $15\mu m$. These will be described in the next chapter, and it will suffice to say here that Becklin's object is quite highly polarised at these wavelengths and

168

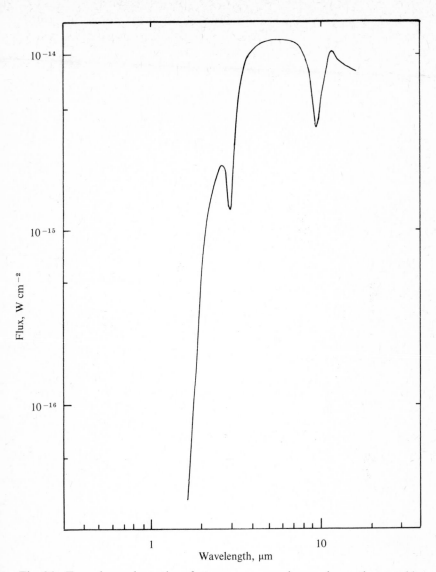

Fig. 36 Two deep absorption features are superimposed on the roughly black-body continuum of Becklin's object. These are due to ice (at 3·2 μm) and silicates (at 10 μm).

that the most credible explanation of this is passage of the radiation through many elongated, aligned grains of dust in the nebula.

So far so good, but a question now raised is how much extinction should there be in front of the **KL** nebula? On the interpretation of Penston *et al*, the **KL** nebula lies very close to Becklin's object and therefore probably experiences much the same reddening. It should therefore have similar polarisation and silicate absorption. But the observations indicate that it has considerably more of both, suggesting an A_v perhaps in excess of 300 magnitudes. Such a value appears inconsistent with the **KL** nebula, being the reradiation cloud surrounding Becklin's object. However, before the heavy-reddening hypothesis is rejected on these grounds the observations should be repeated, for they were made prior to the discovery of the other infrared objects in this region, and it is not now clear just which of the plethora of sources were contributing to the measurements.

Other objections to the model of Penston *et al* have been raised. One of these is based on luminosity arguments.[8] If we apply the heavy-reddening model to all the sources embedded in the 20μm nebula, we must conclude that the energy radiated by them and absorbed by the dust cloud is several million times that radiated by the sun. But the dust must radiate thermally all the energy it absorbs, and Rieke *et al* found the **KL** nebula to have a luminosity of less than 10^5 times that of the sun. (This figure was later raised by a factor of two.)[14] This is probably not a valid argument since the source Rieke *et al* called the **KL** nebula is only the brightest central knot of the extended long-wavelength emission cloud. It is quite possible that most of the luminosity of the entire nebula is provided by the cluster of infrared sources, in which case there is probably not a serious discrepancy between the observed and expected values. If Rieke *et al* are correct, one rather *ad hoc* way out of the inconsistency remains: the infrared sources could be bright stars lying behind instead of within the nebula (much as the Trapezium stars lie on the near side of it) so that most of their radiation escapes leaving only a small portion to be absorbed by the dust.

The other serious objection was published by Becklin, Neugebauer and Wynn-Williams.[15] They measured Becklin's object at 20μm and found it more than one magnitude too bright to be consistent with the heavy-reddening hypothesis. They thus favoured a protostar model in which A_v is small (<10–20 magnitudes) and 600°K dust produces all the infrared radiation.

The controversy has not ended, however: Allen and Penston put up a spirited defence of their model and proposed that the luminous star they favoured has a small infrared excess at 20μm due to local dust.[16] Some evidence supporting this view is provided by the high-resolution spectra of Aitken and Jones.[17] These authors can best fit their data by a model in which Becklin's object has a small silicate bump which is almost obscured by the much stronger silicate absorption which underlies it.

HOW HIGH IS A_V?

Let us distill from this complex discussion the quintessence of the debate. We can state it in the form of a question (which could equally well be asked of quasars but in an entirely different context): how much of the redness of Becklin's object is intrinsic and how much induced?

The observed broad-band energy distribution can be matched fairly well by any one of a range of models. At one extreme the extinction, A_V, is zero and the object is a 600°K black body; at the other A_V is 70 magnitudes and the **BN** source is a star with a surface temperature of 10,000°K or more and a small 20μm excess. In the first case its luminosity is $1000L_\odot$ in the second $10^6 L_\odot$ How do we decide which model to believe?

The third variable, luminosity, gives us no clue for we do not know what proportion of the total luminosity of M 42 to ascribe to Becklin's object. We can place small limits on the range of surface temperature if we use the fact that the 2μm spectrum is featureless.[12] The temperature cannot exceed about 10,000°K otherwise we should see in emission one of the hydrogen lines (Brackett γ) at 2·18μm. Nor is the temperature likely to be in the range 1,500–4,000°K for we should then expect to see the CO and H_2O absorption lines present in the 2μm spectra of most late-type stars (see Chapter 10). And of A_V we can say only that it is certainly not zero, for if it were we should see no ice or silicate extinction. We are left with two distinct models which we can characterise by the value of A_V:

the heavy-reddening model : $65 \leqslant A_V \leqslant 70$ magnitudes

the 'protostar' or dust-cloud model : $10 (?) < A_V \leqslant 45$ magnitude. From the available data on Becklin's star we cannot distinguish these two models. For clues we must also consider some of the other infrared sources which appear to be of a similar nature, and in particular we should try to find the relationship between the value of A_V and the depth of the silicate and ice absorptions. Even if we ascertain such a relationship in one part of the sky we should exercise caution in applying it to others because the relative population of ice and silicate grains on the one hand and absorbing gas and dust on the other is almost certainly not constant throughout the galaxy. In particular we can place little reliance on ice because it is vapourised at quite low temperatures.

Several stars in the Cygnus OB 2 association have A_V's of about 10 magnitudes. Unfortunately they are relatively faint infrared sources, and determining the depths of their absorption bands is difficult. For star no 12, however, there is no trace of absorption due to ice[18] or silicates.[19] Thus we can say that A_V for Becklin's object almost certainly exceeds 10 magnitudes.

THE GALACTIC CENTRE

Becklin's object was not the first source in which an infrared absorption band was recognised. In 1970 Hackwell, Gehrz and Woolf noticed that the 10 and 20μm photometry of the core source at the centre of our galaxy suggested an absorption feature due to silicates.[20] This feature has since been confirmed by Aitken and Jones who found the silicate absorption to be of similar depth to that in Becklin's object.[17] Now, we have a figure for the extinction to the galactic centre, $A_V \sim 27$ magnitudes, deduced from the 2μm colours of the extended source. At first sight, then, this figure gives us an estimate for the A_V of Becklin's object which favours the protostar model. However, this conceals an assumption, namely that the 2μm and 10μm galactic-centre sources are equally extinguished, and in view of their different nature such an assumption cannot be justified. With slightly greater confidence we can set up an inequality in which the 10μm core source is more heavily reddened than the extended 2μm stellar component, and this suggests that for Becklin's object $A_V > 25$ magnitudes.

W3 IRS 5

W3 is a radio-bright section of the H II region IC 1795 in Cassiopeia, and in it a number of discrete infrared sources have been located by scans at 2·2 and 20μm.[21] Of these, by far the most interesting is the one known as IRS 5. Like Becklin's object there is no optical counterpart to this infrared source. The colour temperature is lower – about $250°K$ – and there is a silicate absorption roughly twice as deep as that in Becklin's object.

I cannot escape the conviction that W3 IRS 5 and Becklin's object are fundamentally similar and that the same mechanism is responsible for both their energy distributions. If we try to attribute the colour of IRS 5 entirely to reddening we require A_v to be about 180 magnitudes, roughly two and a half times that of Becklin's object. This is certainly consistent with the relative depths of their silicate absorptions for it implies only a small difference in the silicate grain populations of the two regions. If, however, we adopt the heavy-reddening hypothesis we find that the underlying star in W3 IRS 5, because of its greater distance, must be much more luminous than Becklin's object. So luminous, in fact, that no known star could provide the required energy and we would have to fall back on the sort of mythical object that theoreticians dream of.[22] Furthermore, even if such an object did exist we must face an even greater energy-balance problem than in Orion: the observed far-infrared luminosity of W3 falls short of the energy that must be extracted from the superluminous object by more than an order of magnitude.

So for W3 IRS 5 the heavy-reddening hypothesis is in trouble, and a protostellar model is favoured. By implication, therefore, we must prefer a protostellar explanation of Becklin's object.

NGC 2264 ('DAA6')

But we can counter this argument with an infrared source in Monoceros. NGC 2264 is a bright cluster of young stars including the $4^{m.7}$ O star S Mon. Like Orion it contains a number of T Tau stars, but unlike Orion there is rather little ionised hydrogen and no nebula is visible in small telescopes. An infrared source was found near the southern edge of this cluster independently by Allen at $2\mu m$[1] and by Kleinmann at $10\mu m$ (unpublished). In infrared circles it is often referred to by its pre-publication name DAA 6. This source too has all the characteristics of another of the Becklin's-object class, and in particular it has ice absorption as deep and silicate absorption nearly half as deep as that in Becklin's object.[23] But DAA 6 is not without an optical counterpart, as plate 17 shows. It lies at the tip of a small fan of nebulosity which spectra show to be a reflection nebula.[2,24] When we ask which star is illuminating this nebula, we find none of the visible stars can do so. The infrared source must therefore be responsible, and this implies that it is a normal star. Thus, to produce the observed infrared colours, it must experience an A_v of about 40 magnitudes in our direction, but much less in the direction of the reflection nebula: it is embedded in a dense but patchy dust cloud. Were the dust removed, DAA 6 would be about

17. NGC 2264. The infrared source DAA 6 is at the apex of the nebula arrowed. The dramatic dark nebula below centre is the Cone Nebula.

one thousand times as bright as the brightest star on plate 17, and of apparent magnitude near zero.

The energy distribution of DAA 6 shows a steep rise at 20μm, suggestive of the **KL** nebula. But at 100μm no signal has been detected. As with other sources there is a discrepancy between the supposed absorbed and reradiated fluxes; however, the discrepancy is not large in this case. The source should be detectable at 100μm if a small improvement in sensitivity can be made.

Again resembling Becklin's object, DAA 6 lies in a molecular cloud. This one was discovered quite by chance, for the intention of Zuckerman and his collaborators was to search for molecular emission from the head of the Cone Nebula, to the south of DAA 6 in the plate. Due to an error in the adopted coordinates, the radio telescope was set exactly on to DAA 6, and strong line emission was recorded. What made this coincidence particularly fortuitous was that the radio astronomers made this observation before the discovery of the infrared source had been announced. The cloud is several minutes of arc in diameter, and at optical wavelengths is quite invisible except for the small section illuminated by the infrared source.

It is perhaps worth noting *en passant* that the Cone Nebula comprises a number of linear-bright rims interrupted by denser globules. If the reader cares to extend the linear sections on plate 17, he will find that they converge on the infrared source. The author, who is rather attached to this source, feels that this is trying to tell him something; much of the astronomical community doesn't agree with him.

If we use DAA 6 as our yardstick, we must favour the heavy-reddening hypothesis. Combining all our present information only clouds the picture, and in all fairness it must be said that the issue is nowhere near a solution. Probably the answer lies in finding more of these infrared sources and in investigating the depth of their silicate and ice features as a function of the possible values of A_V. Quite a large number of potentially interesting objects were recorded in the Air Force survey, and others have been found by the sort of searching described at the beginning of this chapter. Below is a brief list of some of the infrared sources which may prove to be relevant to this discussion when further observations are made.

USAF Source in Cygnus

This isolated source, described by Merrill and Soifer, closely resembles

Becklin's object in its energy distribution and has rather weaker ice and silicate absorption bands.[25] Its distance is not known and can only be inferred from the radial velocity of the associated molecular emission. If this figure is used, the heavy-reddening model would imply an embarrassingly large luminosity for it.

Rosette Source

Recorded in the USAF survey and identified by Cohen, this infrared source lies in the Rosette Nebula (NGC 2237–9).[26] Very close to it is a fan-shaped nebula whose nature has not been determined. The energy distribution of the Rosette source is rather different from that of Becklin's star or DAA 6, and if reddening is responsible the local reddening law is quite abnormal.

Orion Sources

The other sources in the cluster around Becklin's object appear to obey a relationship in which the redder the object, the deeper the silicate absorption.[8,16] This supports the heavy-reddening hypothesis, but it should be made clear that the observations are not yet of sufficient quality to place much reliance on such a correlation. A similar result seems to be implied by Borgman's observations of the galactic centre sources.[27]

H II Regions

The Westerhout catalogue of H II regions makes good hunting for infrared astronomers. In addition to W3 IRS 5, sources have been found in W49,[28] W51 and W75.[29] Others were found in NGC 7538,[29] Sharpless 228[30] and 269[29], both small H II regions, and yet another in M17.[31] All these should be examined for absorption bands.

G 333·6–0·2, the very luminous H II region mentioned in Chapter 7, has been studied by Aitken and Jones who maintain that there is strong silicate emission and absorption almost exactly cancelling out.[32] But if they use the silicate-A_V calibration implied by the heavy-reddening hypothesis for Becklin's object, they deduce a luminosity for the H II region which is disturbingly high.

IC 430 = Haro 13a

This peculiar nebula near 49 Ori contains a star which can be recorded photographically in the near infrared. At longer wavelengths the energy distribution is rather peculiar (Fig 37) and can best be explained by the effects of reddening on a free-free continuum.[33] If this interpretation is correct, A_V is about 25 magnitudes, and this is

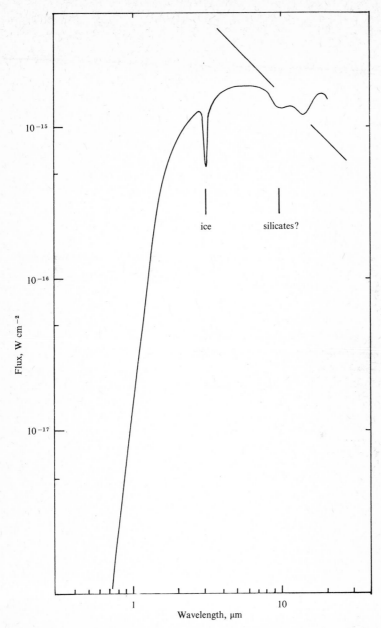

Fig. 37 The energy distribution of Haro 13a (1C 430) also shows absorption due to ice and, possibly, silicates. Apart from a 20μm excess, the energy distribution can be synthesised by the effects of reddening on a free-free continuum.

the most heavily reddened star known which can be recorded photographically. It is therefore of interest to search for ice and silicate absorptions in it. As Fig 37 shows, there is quite definitely ice absorption, and there may be a hint of silicates too. If we use this as the calibration of Becklin's object, we must come out on the side of the heavy-reddening model.

OH 0739–14

At these coordinates in Puppis lies an infrared source which is almost certainly associated with an OH emission region.[29] The source is extended in the infrared and its colour is not very red, corresponding to a hot object reddened by $A_v \sim 15$ magnitudes. In addition, there is a silicate absorption at least as deep as that in Becklin's object. So this piece of evidence favours the protostar model.

THE INFRARED SKY

So the infrared sky is populated not only by all the objects optical astronomers observe, but also by true infrared sources whose nature remains enigmatic. They may be the brightest stars in our galaxy; they may be the youngest. They are certainly associated with dark molecular clouds whose very existence was until recently quite unknown, and whose number may yet prove to be large. As an example, strong molecular emission from the Ophiuchus dark cloud was found only after the infrared sources had been reported.[34] Progress in both infrared and microwave studies will probably proceed hand in hand.

We are still exploring the infrared sky, especially in the southern hemisphere, and can by no means claim to know it. It is with much trepidation, therefore, that the author has prepared Figs 38–40 which compare the sky in the visible and at 2·2 and 10 μm. Particularly south of $-30°$ the infrared maps will be very incomplete. Also it should be noted that most of the 10 μm sources are extended, so how bright they appear is a function of the size of the beam employed. But the maps will at least give the reader an idea of how different the sky would be had he infrared-sensitive eyes. Tables 5 and 6 list the ten brightest sources at 2·2 and 10 μm.

TABLE 5
The Brightest Stars at 2·2 μm

Star		K magnitude	
α	Ori	−4·5	
R	Dor	−3·9	
α	Sco	−3·8	
α	Her	−3·4	
o	Cet	−3·3	at maximum
β	Gru	−3·2	
γ	Cru	−3·2	
W	Hya	−3·1	
α	Boo	−3·0	
R	Leo	−3·0	at maximum

TABLE 6
The Brightest Objects at 10 μm

Source	N magnitude	
η Car	−7·8	
IRC+10 216	−7·2	
Messier 17	−6·3	extended
VY Cma	−6·0	
NML Cygni	−5·4	
NGC 6357	−5·4	very extended
α Ori	−5·2	
CIT 6	−5·1	
o Cet	−5·1	
Galactic Centre	−5·1	very extended

179

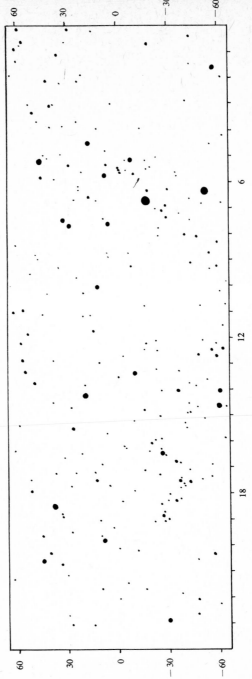

Fig. 38 The sky in the visible between $\pm 65°$ declination. There is some distortion in high declinations on this projection.

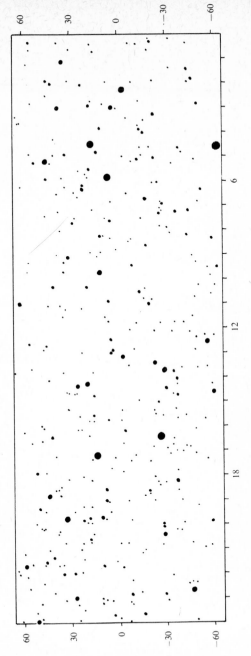

Fig. 39   The sky at 2·2 μm. The constellations are no longer recognisable, but the bright red stars such as Aldeberan, Betelgeuse, Arcturus and Antares are prominent.

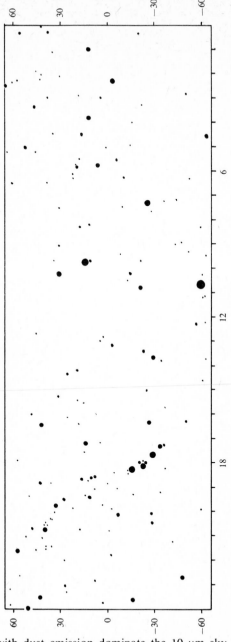

Fig. 40   Objects with dust emission dominate the 10 μm sky. Some of these sources are extended.

10
DEVELOPMENT OF THE SCIENCE

The six observational chapters of this book have concerned themselves almost exclusively with photometry – the simple measurement of the energy radiated by celestial sources in various selected wavebands. But photometry is only our first step in attempting to unravel the universe. Had we at optical wavelengths not progressed beyond simple photometry, our understanding of astronomy would be much more circumscribed than is now the case. Paramount among the more sophisticated techniques which have contributed at optical wavelengths is spectroscopy; polarimetry and sky-limited photography have also proved themselves of great value. In the same way we can hope to broaden greatly our knowledge by applying these techniques at infrared wavelengths. This, of course, is easy to write, but tough to put into practice. We have, after all, made rather heavy weather of the photometry and can claim to have mastered it only in the last decade or less. Since we have to be able to walk before embarking on greater things, the consequent disciplines of infrared spectroscopy and polarimetry are yet in their infancy. Hence their relegation to the closing act of this book.

SPECTROSCOPY

If black bodies were the only constituents of infrared energy distributions, we could already claim to have discussed spectroscopy, for our photometry in the atmospheric windows is sufficient to resolve a black body and determine its colour temperature and luminosity. We have found, however, that superimposed on the crude black-body distributions are narrow emission and absorption bands, in particular the 10 μm silicate feature. It has therefore become important to use narrower wavebands, defined by interference filters, which allow us to resolve the silicate profile. But with each increase in spectral resolution we find new features which demand still narrower wavebands for their study. Progressively, narrow-band photometry becomes spectroscopy, and just where we draw the line is a matter more of semantics than scientific reason. For the purposes of this book it has been convenient to introduce the ice and silicate bands early into the flow and to retain for the present chapter the narrower spectral features whose existence would not normally be demonstrated by broad-band photometry.

Instrumentation

There are several ways of producing spectra at infrared wavelengths. The simplest of these is merely an extension of narrow-band photometry by the use of extremely narrow interference filters. But it becomes impracticable to mount in a dewar the large number of filters needed to span, say, the 10 μm window at high resolution, and in any case the cost of such an approach is prohibitive. Instead, a continuous filter is employed, one in which the wavelength of peak transmission varies with position on the filter. The continuous filter is usually arranged in the form of a circle or semicircle mounted in a filter wheel: the wavelength of maximum transmission varies continuously around the circle, say from 7·5 to 15 μm. The resolution attainable with such a system is of order 0·1 μm. The filter wheel may be rotated to a position corresponding to any chosen wavelength within the available range, and its rotation would normally be controlled by a digital stepping motor. A filter-wheel spectrometer for the 10 μm window was built by Gillett and is frequently used at Mount Lemmon and Kitt Peak.

A spectrometer which to optical astronomers might appear more conventional, was constructed by Aitken of University College, London. He uses a reflection grating to produce a spectrum, and by tilting the grating can cause any chosen wavelength to fall onto the detector. The grating spectrometer is capable of slightly greater spectral resolution than the filter-wheel spectrometer and has essentially the same sensitivity for an identical detector.

Either of these systems is abysmally inefficient because the detector examines only one wavelength at a time. By comparison, a spectroscope operating in the optical would employ a photographic emulsion or the photocathode of an image tube as its detector. One- or two-dimensional detectors such as these can integrate on all wavelengths simultaneously, thereby gaining a factor in observing time equal to the number of wavelengths to be sampled. For a 10 μm spectrometer this factor is typically of order 100.

In principle we can regain the missing factor by using a Michelson interferometer. In this clever but complicated device all the wave-

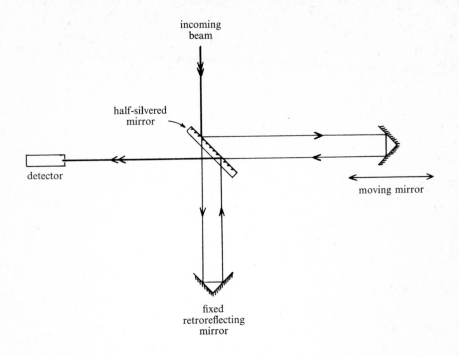

incoming
beam

half-silvered
mirror

detector

moving mirror

fixed
retroreflecting
mirror

Fig. 41 Schematic drawing of the Michelson interferometer. As the right-hand retroreflecting mirror ('cat's eye') is moved the signal received by the detector is modulated.

lengths of interest are simultaneously received by the detector, but each wavelength is coded by chopping it at a different frequency. This resulting signal falling on the detector is rather jumbled, but by the mathematical technique of Fourier transformation it can be decoded to yield the incident energy at each wavelength.

The chopping (or modulation) is no longer performed by a rotating or vibrating mirror as described in Chapter 2. Instead, the principle of interference is utilised. The incoming radiation is split into two beams both of which are reflected off mirrors before they recombine, as Fig 41 shows. Now consider radiation of one wavelength. If the two paths traced by the radiation differ in length by an integral number of wavelengths, the radiation will interfere constructively and a signal will reach the detector. If, however, there is an odd half wavelength difference in path length, the interference will be destructive and no signal will arrive. In the Michelson interferometer one of the mirrors is continuously moved so as to vary the path difference, and the signal from a monochromatic source therefore varies sinusoidally in amplitude. The frequency at which this modulation occurs is determined in part by the velocity of the moving mirror, which we can measure, and in part by the wavelength of the radiation. The longer the wavelength, the slower the modulation. Thus when radiation of a broad waveband is allowed into the interferometer, each wavelength will be modulated at a frequency inversely proportional to that wavelength.

The Michelson interferometer is capable of much higher spectral resolution than the grating or filter-wheel spectrometers. This fact, together with its greater efficiency and hence higher signal/noise ratio, makes it deservedly popular for near infrared spectroscopy. It is however, a very difficult piece of equipment to make work at 10 μm where all its components radiate strongly. Most of the 10 μm spectra which have been published to date were secured with filter-wheel or grating spectrometers.

1–5μm; the Molecular Bands

It is in the PbS region that the best infrared spectra of astronomical objects have been taken, for at these wavelengths Michelson interferometry is relatively easy and can be pursued to very high resolution indeed. Moreover, there are a great many spectral features to be investigated, most of them arising from diatomic and polyatomic molecules. Of course the best-known molecular absorption bands in the PbS region are those due to CO_2 and H_2O, those villains of the earth's atmosphere which cause infrared astronomers so much heartache. At high resolution these bands can be split up into a vast number of extremely narrow absorption lines all roughly evenly spaced. To find the reasons for this we must delve into molecular spectroscopy, a subject quite sufficiently complicated to fill this entire book. It is possible to shortcut most of the complexities by noting that gaseous molecules behave rather like solid balls connected by springs. They can be set into vibration if they are hit by photons of the right frequency, and in many cases the right frequency occurs in the infrared. The vibrations can occur in a number of different modes including tensional (back and forth along the spring), torsional (coiling and uncoiling of the spring) and lateral (bending of the spring). Combinations of these different modes produce the fine structure lines. Fig 42 illustrates the detail present in a molecular spectrum.

At the edges of the atmospheric absorption bands it is possible, although difficult, to distinguish lines of CO_2 in astronomical sources.

The difficulty is lessened if the source has a large radial velocity, for the lines of the source are then shifted by the Doppler effect away from the atmospheric features. Water can also be detected because at high temperatures it has a slightly different spectrum from that of the cooler water vapour in our atmosphere. Steam absorption bands occur at $1\cdot9\,\mu$m. Other molecules with lines in the PbS region include OH, CO and CN.

The finest work is undoubtedly that of the French team Pierre and Janine Connes who have received many accolades for their spectra of the planets. An atlas of the 1–$3\,\mu$m spectra of Venus, Mars, Jupiter and Saturn has been complied by the Connes,[1] and some isolated papers have also been published. These include an announcement of the discovery of HCl and HF in Venus' atmosphere,[2] and the determination of the CO/CO_2 ratio ($4\cdot5\times10^{-5}$), also in Venus.[3]

High-resolution spectroscopy of the planets has also been practiced by Beer and others. Among their work in the 3–$4\,\mu$m window is measurement of the methane isotope CH_3D abundance in Jupiter.[4] Here one of the hydrogen atoms has been replaced by deuterium (heavy hydrogen), and Beer *et al* were able to measure the deuterium/hydrogen ratio in Jupiter. They found a value much lower than that on earth. In another paper Beer, Norton and Martonchik examined the 3–$4\,\mu$m spectrum of Venus and found a broad absorption feature which might be due to bicarbonates.[5]

Fig. 42 The fine-structure lines in a molecular spectrum. Carbon monoxide in the atmosphere of Venus produced this regular series of absorption lines. The three vertical bars identify terrestrial atmospheric lines. This diagram covers approximately 0·0044 μm, about one hundredth of the width of the 2·2 μm window.

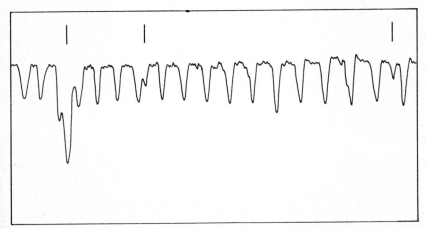

Recently the Connes, in pushing to higher spectral resolution, have also turned their attention to more distant sources.[6] Beyond the solar system it is undoubtedly the late-type stars which are of special interest in the PbS region because the atmospheres of these stars are cool enough for molecules to populate them. Most of the early results in this field originated with Johnson, Thompson and others at the University of Arizona. Johnson's first valuable contribution was to spectral type NML Cygni on the basis of the strength of its CO absorption bands.[7] He also found a broad feature at $3\cdot1$ μm which he attributed to ice. The same feature was found by Gillett, Stein and Low who used a filter-wheel spectrometer to cover the $2\cdot8$–$5\cdot6$ μm region.[8] These authors postulated ammonia as an alternative to ice, but it now seems that ammonia is not the cause of the feature.

In 1969 and 1970 a string of papers presented identifications of the features in the PbS region then found in late-type stars.[9] Among the molecules identified were CO and H_2O (steam), C_2, CH, CN and TiO. Steam bands occur principally in Mira variables and, according to Frogel,[10] correlate well with the H_2O emission line at $1\cdot35$ cm. OH bands are present in the 3–4 μm spectrum of a Ori.[11]

The relative abundances of the isotopes of carbon and oxygen can be determined from these spectra, for the absorption lines are shifted in wavelength if one of the constituents has a different atomic weight. In this way the carbon 12 to carbon 13 ratio may be determined. In K and M stars the C_{12}/C_{13} ratio is as low as 4 or 5, but in carbon stars it is higher by a factor of three. This may suggest the continuing manufacture of C_{12} in carbon stars. The relative abundances of O_{16} and O_{17} are best measured in the 5 μm region where very strong CO lines are found.[12] The depression of the continuum at 5 μm by these lines can be detected by broad-band photometry. Rank, Geballe and Wollman found the ratio O_{16}/O_{17} in IRC+10 216 to be about 400.[13]

The year 1972 witnessed the advent of higher resolution work in the 1–2·5 μm windows.[14] In addition to the molecular species already mentioned, these spectra revealed the presence of a great number of narrow absorption lines due to neutral metals.

The 10 μm *Region*

The only identified features at 10 μm which can be attributed to molecules are those due to silicon carbide and silicates. As for simpler molecules such as CO, it should be possible to resolve the silicate feature into fine structure lines. That this has not proved so may be for one of two reasons, and probably both: either the material is too coarse or it is not a single, pure silicate crystal but some amorphous mixture of the silicates of iron, magnesium, aluminium etc.

Spectroscopy at 10 μm is not just concerned with molecules, however. When Johnson set up the infrared photometric bands it was believed that the most important feature at 10 μm would prove to be an emission line at 12·8 μm due to singly ionised neon. The 10 μm window was consequently allocated two bands, O and P, only the latter of which contained this line. By securing photometry in both bands it was anticipated that the strength of the Ne II line could be investigated. This has never been done, principally because of the contamination of such photometry by the presence of circumstellar dust, especially silicates. In any case, no source has yet been found in which the 12·8 μm neon line is strong enough to influence significantly broad-band photometry.

The first detection of Ne II, in the planetary nebula IC 418, was made by Gillett and Stein in 1969.[15] Few other sources have yet been found to contain measureable Ne II emission. Indeed it was as

Fig. 43 The 10 μm spectrum of NGC 7027. Of the many features in this spectrum only the S IV and Ar III emission lines have been identified. The portion between 9 and 12 μm has been examined at higher spectral resolution than the outer sections.

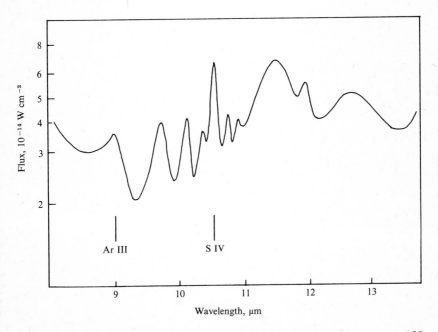

recently as 1974 that a Ne II line stronger than that in IC 418 was discovered, in the H II region G333·6–0·2.[16]

Between 1970 and 1973 lines of Ar III at 9·0 μm and S IV at 10·5 μm were found in the plantary nebulae NGC 6572 and 7009, and in the peculiar emission nebula NGC 7027.[17] This last-named nebula has already been described in Chapter 7 and is shown in the plate, p 136. It is a particularly interesting object at both visible and infrared wavelengths in addition to being a bright radio source. Although variously classified as a planetary nebula and (as in Chapter 7) a compact H II region, NGC 7027 is one of those objects which defies classification. Among all planetary nebulae and possible planetary nebulae it is the brightest at 10 μm, and is therefore an excellent subject for infrared spectroscopy. In 1973 two papers were published on the 10 μm spectrum of NGC 7027 – one by the San Diego group,[18] the other by the University College London group.[19] Fig 43 combines these two spectra and indicates the features identified. The most puzzling feature is the broad band at 11·3 μm: a suggested origin for this is carbonate material, but more justification is needed before this can be accepted. When higher-resolution spectra are eventually secured, the 11·3 μm band may well be resolved into a cluster of single emission lines.

POLARIMETRY

Our everyday world contains a considerable amount of polarised light, as anybody who has experimented with a pair of polaroid sun-glasses will know. Polarisation occurs whenever light is scattered or reflected obliquely off a surface. In the astronomical environment polarisation is also induced by scattering, but in this case there are no extended surfaces off which light can reflect, and the only available solids are those tiny grains of interstellar and circumstellar dust we have encountered so often since Chapter 4.

The scattering of starlight through a uniform cloud of spherical dust grains cannot induce polarisation. Each reflection occurs in a different plane determined by its location on the grain. A collection of randomly oriented reflections merely cancel to give no nett polarisation. To give rise to a significant polarisation one of two conditions must be fulfilled. Either the grains are elongated and have their axes aligned so that reflections occur principally along a plane parallel to the alignment, or the grains are spherical but lie in a very flattened distribution approaching that of a disc. Magnetic fields can align grains, so polarisation at optical wavelengths can sometimes be used to trace the presence and direction of magnetic fields, as in the spiral arms of our galaxy.

But this is not all. Some information on the nature of the dust grains themselves can be deduced from measurements of the degree of polarisation as a function of wavelength. It is on this count in particular that there is value in determining the polarisation at infrared wavelengths, for by combining optical and infrared data we can hope to narrow down the range of possible grain compositions.

The principle involved in measuring polarisations is simple. In the light path of our photometer we place a polarising filter (analyser) which can be rotated into various positions. For use at PbS wavelengths the analyser is usually made of plastic, but at 10 μm fine-mesh wire-grid analysers are usually used. Neither of these is as efficient as optical polarising filters in that they allow through some radiation of all angles of polarisation. Two methods are employed. In the simpler, normal photometry is performed with the analyser in a number of different orientations. From the relative signals the polarisation may be deduced. A more complicated method abandons the square-wave chopping between star and sky. Instead the analyser is continuously rotated. A modulation of the signal falling on the detector occurs only if the source is polarised.

There are two parameters to be measured. One of these is the position angle on the sky in which the radiation is polarised; polarimetry is unique in its ability to indicate a preferred direction. The other is the degree of polarisation, p, usually expressed as a percentage. If the signal in the position angle of maximum polarisation is S_{max} and the corresponding minimum signal 90° out of phase is S_{min}, p (as a percentage) is defined by:

$$p = 100 \frac{S_{max} - S_{min}}{S_{max} + S_{min}}$$

At infrared wavelengths, polarisations are small. Values of p in excess of 1 per cent are rare, and this means that a photometer and telescope capable of detecting a star of 10th magnitude at K will be hard put to measure a polarisation in a star fainter than 4 or 5 at K. It is for this reason that most infrared polarisation measurements, especially at 10 μm, have been made of the brighter late-type stars in the IRC.

The first infrared polarisation measurements were made by Forbes as long ago as 1967.[20] Since at that time the only 'infrared stars' known were NML Cyg and NML Tau, these were the sources he measured. At 2·2 μm Forbes found NML Cyg to have a polarisation of about 3 per cent; he could not have anticipated that it would be six years before a higher value was found at this wavelength. Other stars with quite large polarisations were quickly discovered, notably VY CMa[21] and IRC+10 216[22] in 1970. Russian observers, too, were initially active in this field.[23]

A comprehensive summary of the available polarisation measurements was compiled by Dyck, Forbes and Shawl in 1971.[24] At that time the largest known 2·2 μm polarisation was still that of NML Cygni, although other sources had much larger values at shorter wavelengths. In particular IRC+10 216 had been found to be 20 per cent polarised at 1 μm.[22] It will be recalled that this source is a small elliptical object on deep optical photographs; the direction of polarisation was found to be along the minor axis of the ellipse.

Dyck *et al*'s compendium contained only late-type stars from the IRC and included no measurements at wavelengths longer than 3·5 μm. The principal developments since 1971 have been towards longer wavelengths and different types of infrared source. A few 10μm polarisations were published by Capps and Dyck in 1972, most of them being rather small.[25] Of the other infrared sources investigated, by far the largest polarisations were found for those enigmatic sources described in the last chapter which show evidence of heavy reddening. Becklin's object was the first of these. At 2·2 μm it has the largest polarisation yet measured, about 15 per cent, or more than four times that of NML Cyg.[26,27] This large value was discovered independently by three groups, in order: Dyck at Kitt Peak, Loer and Allen at Mount Wilson, and Breger and Hardorp who also observed from Kitt Peak. So large a value of the polarisation is thought to arise only by scattering of the radiation through very well-aligned particles. Thus we conclude that Becklin's object is quite heavily reddened and that the Orion complex contains elongated dust grains aligned probably by a magnetic field. The position angle of polarisation is about 110°–290°, but this does not correlate with any optical feature, as indeed we should expect knowing the dust to be in the dark molecular cloud. These conclusions are strengthened by polarisation measurements at longer wavelengths.[28] From 2 to 8 μm the polarisation falls steadily; this is the normal trend throughout the infrared and, like the corresponding fall in extinction, is caused by the progressive increase in the ratio of the wavelength of the radiation to the dimensions of the dust grains with which it interacts. From 8 to 13 μm Becklin's object has a deep absorption due to silicates. The polarisation through this band varies as the extinction, reaching a maximum of about 10 per cent at 10 μm.

The galactic centre core source has a similar polarisation at the shorter wavelengths,[29] but has yet to be measured through the silicate band. The reddened source in NGC 2264 has a smaller polarisation of about 2·5 per cent at K.[26] It therefore seems likely that when measurements are made, many reddened sources will be found to have very large polarisations. Some (eg W3 IRS 5) may even exceed Becklin's object. Thus Table 7, which lists the largest 2·2 μm polarisations known at the time of writing, may very soon be rendered out of date.

TABLE 7

The largest 2·2 μm polarisations

Object	Polar-isation (%)	Position angle (°)	
Galactic centre (core source)	15 ?		{ Not measured at K, 8% polarised at L
Becklin's object	14	120	7% polarised at L
NML Cygni	3·3	80	
VY CMa	2·8	60	
DAA 6 (in NGC 2264)	2·7	110	
NU Ori	1·9	160	in Messier 43
IRC+30 021	1·7	80	
IRC+10 216	1·6	120	
L$_2$ Pup	1·2	140	
CIT 6	1·2	10	
VX Sgr	1·0	150	
IRC+20 052	1·0	90	

What can we learn from polarisation measurements in the infrared? As yet rather little, but this is mostly because our sensitivity is still poor and the existing data are few and of rather low accuracy. Some interesting results have emerged. For example, a good correlation exists between the presence of an infrared excess and of visible and near-infrared polarisation.[30] This correlation is exactly what we expect, since both the polarisation and the infrared excess are caused by circumstellar dust. It is, however, comforting to find such confirmation, for in the field of astronomical research the best-laid theories have a habit of being laid low by observations.

One of the most valuable uses of polarimetry from optical to infrared wavelengths is in helping to identify the composition of the dust grains causing the scattering, or at any rate in serving to eliminate some of the possibilities. Each type of grain gives rise to a characteristic wavelength-dependence of the polarisation. If data of sufficient accuracy can be taken, the grains ideally can be identified. In practice, the various grain mixtures which might be present give rise to very similar polarisations. Perhaps the best work along these lines was that of Gehrels who examined the polarisation of the reddened star ζ Oph from the extreme ultraviolet to 2·2 μm. Yet Gehrels could say only that the polarisation was consistent with that expected from the sort of grains theoreticians select to populate interstellar space, namely graphite particles coated with shells of ice.[31] Similarly, Dyck, Forbes and Shawl could make only qualitative arguments that the polarisation of VY CMa might be produced by silicate grains.[24] Although this is potentially a rewarding line of research, we clearly have a long way to go.

INFRARED PHOTOGRAPHY

It was categorically stated at the start of Chapter 2 that photography beyond 1 μm was impossible. For the purposes of this book that statement is correct, but in the closing pages I should add a few words to redress the balance, for the statement is no longer true.

The ability to image a scene, and hence take photographs, at infrared wavelengths is of obvious value in military applications: men and internal-combustion engines can both be made visible at night if we survey them at 10 μm. Devices which would accomplish just this have been developed in various top-secret institutions in England, the USA and probably elsewhere. In the last few years infrared imaging systems – *vidicons* – have been released to the commercial market. Although at first sight vidicons are insufficiently sensitive for useful astronomical applications, it is by no means clear that this will always be the case. The late 1970s will probably see the publication of infrared photographs of planets, comets, H II regions, molecular clouds and maybe even galaxies. And just what we will then find cannot be guessed, for infrared astronomy is a field in which predictions are all too often proved wrong.

11

PROSPECTS

The ardent collector of Golden Ages could probably catalogue several in the field of astronomy. Of these one should certainly be claimed by the decade beginning about 1960. It was during this period that quasars became known and pulsars were discovered. It was at this time too that the opening of the electromagnetic spectrum was made possible. Of the new astronomies that emerged in these years, cosmic ray, X-ray, ultraviolet, infrared and microwave, it is the infrared which has made the greatest impact on the science. It is natural to ask whether infrared astronomy is merely a precocious upstart now nearing the end of its hour of glory or whether, like radio astronomy, it will establish itself as an important lasting contributor.

Any specific type of astronomical research must eventually grind to a halt because the number of objects to which it can be applied is finite. Research progresses in versatility. In some cases the time scale for depletion of the sky is long; an example of this is abundance analysis by high-resolution optical spectroscopy. Other branches run their courses more rapidly, and infrared photometry as currently practised is one of these. The longer wavelengths will run out first; PbS photometry can be applied to a sufficiently large number of stars that it will be some years before all the useful information is extracted. But limited they both are. Any lasting contribution must therefore be made either by the development of much more sensitive detectors allowing a vast increase in the number of available sources, or in the field of infrared spectroscopy. As I write this, a few observatories in the USA are attempting to develop indium antimonide detectors for the $1-5\,\mu$m region. These detectors are temperamental, the electronics are complicated and both the detectors and their pre-amplifiers must be cooled at least by solid nitrogen. Preliminary results are encouraging and suggest that InSb is capable of an order of magnitude higher sensitivity than PbS cells for $1-4\,\mu$m and an even greater improvement at $5\,\mu$m over doped Ge detectors. But continued development of this sort cannot be guaranteed and we must place our reliance on spectroscopy. Since 10 μm spectroscopy is itself limited to a rather small number of sources, it seems that the future lies in the PbS region. When satellites can carry our telescopes above the earth's atmosphere, molecular spectroscopy will be able to tell us a great deal about the conditions in cool stars; as a result there will be valuable work to be performed in investigating the post–main–sequence stages of stellar evolution. This, then, is the best prospect for infrared astronomy.

This is a deliberately pessimistic view, for it assumes no versatility to infrared astronomy and assumes that no major discoveries of new objects will be made, discoveries equivalent to those of quasars and pulsars. So it would probably not be sticking my neck out too far to hazard the opinion that infrared will still be making significant contributions to astronomy by the end of the century.

Infrared will not stand alone in the way that optical and, for a time, radio astronomy have done. In part this is an historical accident: infrared came too late onto the scene to be independent of the others. In part too it results from the wholesale availability of all spectral regions. We have found that we can better broaden our understanding if we study single objects at many wavelengths rather than many objects at one wavelength. And in this respect it is no longer meaningful to ask whether infrared has a future in the way that radio astronomy has had, As long as there are fresh objects to study there will be a demand for infrared observations of them, The sort of questions that might be asked, and that infrared can best answer, are:

Does the object have circumstellar dust clouds and of what sort?

Is there a cool star associated with the object? What sort of star is it?

Is there ionised gas producing free-free emission?

How heavily reddened is the object?

This book has described the age of discovery in infrared astronomy. For ten years now we have been like Herschels finding new and exciting things everywhere we point our telescopes. But this age is drawing to a close. Most of the discoveries have now been made and the whole business of infrared astronomy is slowing down to a respectable pace in which routine, more exacting and perhaps rather mundane observing occupies most of one's time at the telescope. If books dedicated solely to infrared astronomy are written ten years hence they will read very differently from this one.

This has undoubtedly been an exciting decade. A small number of astronomers, many of whom (like the author) just happened to be in the right place at the right time, have been able to cash in on the discoveries and write their names at the heads of a large number of scientific papers. Since 1964 infrared astronomy has accumulated an enormous literature, about two thirds of which is included in the references to this book. The rate of production of papers will probably drop; indeed there is already some indication that it is doing so. The next generation of papers should be longer and more carefully written and they should eschew the superficial treatment of the subject commonly used to date (and hence imposed on this book). This decade has led to an information explosion in infrared astronomy. During the next decade we will sit down and work out what it all means and how it relates to the information garnered at optical, radio and other wavelengths. We are at a turning point.

APPENDIX I

THE PLANCK FUNCTION

The theoretical derivation of Planck's radiation function is beyond the scope of this book, and the mathematical equation governing it must be given without explanation. There are a number of ways of expressing it; the one used here is:

$$F = \frac{c_1}{\lambda^4} \, \varepsilon(\lambda) \, \frac{1}{e^{c_2/\lambda T} - 1} \qquad\qquad 1$$

The equation is not so fearsome as it at first sight seems. F is an energy flux radiated by the object and received by a detector. The units employed here are Watts cm^{-2}. This mixture of c.g.s. and M.K.S. units may offend the purist. Astronomers invariably express numbers in the most bizarre of units. For example, Hubble's constant, which is a measure of the age of the universe, is usually quoted in km s^{-1} megaparsec $^{-1}$.

λ is the wavelength employed, measured in μm. $\varepsilon(\lambda)$ is the emissivity, a function of λ and a property of the body under observation which at optical wavelengths we could call its colour. Henceforth in this appendix we shall put $\varepsilon(\lambda) = 1$: this defines a black body. Two constants remain: c_1 and c_2. c_1 is a scaling factor given by:
$$c_1 = 2\pi hc^2; \qquad c_2 = hc/k$$
Here h is Planck's constant, c is the velocity of light and k is Boltzmann's constant. c_2 has a value 14,388 μm K.

F, defined as above, is the monochromatic energy flux – the flux at a given wavelength from an object of temperature T. It is also proportional to the number of photons of given wavelength emitted by the object. Plots of F against λ for several values of T will be found in Figs. 3 and 4, chapter 1. In practical applications we cannot measure a monochromatic flux, but must make our measurement over a small range of wavelengths, $\delta\lambda$. We thus need to know how much flux is radiated within our bandpass, and this is given by the formula

$$F_\lambda = \frac{c_1}{\lambda^5}\, \frac{1}{e^{c_2/\lambda T} - 1} \qquad\qquad 2$$

F_λ is measured in Watts cm $^{-2}$ μm^{-1}, and the flux we measure is therefore $F_\lambda\, \delta\lambda$, or

$$F_{measured} = \frac{c_1}{\lambda^5}\, \frac{\delta\lambda}{e^{c_2/\lambda T} - 1} \qquad\qquad 3$$

One final form sometimes met is F_υ. If we are concerned with the energy radiated within a frequency band, δv, we have:

$$F_{measured} = \frac{2\pi h}{c}\, \frac{v^3\ \delta v}{e^{hv/kT} - 1} \qquad\qquad 4$$

where we have substituted for c_1 and c_2. F_v is usually measured in W m^{-2} Hz^{-1} or, since one Hertz is a minuscule portion of the spectrum, in flux units (now Janskys) such that 10^{26} flux units equals one W m^{-2} Hz^{-1}.

The forms F_λ and F_v have been introduced here because many astronomical papers use these in preference to the monochromatic flux F. They are inter-related by $F = \lambda F_\lambda = v F_v$; indeed where F is used in the literature it is usually written λF_λ. Physics texts are generally written in terms of F_λ, which is then often called $B_\lambda(T)$, the black body function of wavelength and temperature.

The Rayleigh-Jeans approximation
The Planck function is sharply peaked, reaching its highest at $\lambda = \lambda_{max}$. If λ is very large, $c_2/\lambda T$ is tiny and we can simplify equation 1 by writing
$$e^{c_2/\lambda T} \sim 1 - c_2/\lambda T. \quad \text{Equation 1 then becomes:}$$
$$F \propto T\lambda^{-3} \qquad\qquad 5$$
This is the Rayleigh-Jeans approximation and is valid when $\lambda \gg \lambda_{max}$. On logarithmic axes (as in Fig. 4) this is a straight line of gradient -3. It is usually referred to as the Rayleigh-Jeans tail of the Planck function.

Stefan's law
Stefan's law relates to the total energy radiated by an object over all wavelengths. To derive this we must integrate equation 3 (or 4) over all wavelengths and multiply it by the area, a, of the body.

Total energy, $E = \int_0^\infty \dfrac{c_1\, a}{\lambda^5} \; \dfrac{d\lambda}{e^{c_2/\lambda T} - 1}$ Watts

The integral is best manipulated by putting $x = c_2/\lambda T$. It then becomes

$$\frac{a\,T^4}{c_2^{\,4}} \int_0^\infty \frac{x^3 dx}{e^x - 1}$$

which integrates to:

$$\frac{\pi^4}{15}\, a\, \frac{T^4}{c_2^{\,4}}$$

We therefore have $E \propto a\,T^4$, which is Stefan's law. The constant of proportionality, Stefan's constant σ, is given by:

$$\sigma = \frac{2\pi^5 k^4}{15 c^2 h^3}$$

In full, Stefan's law also contains the factor ε (the emissivity).

Wien's law

Wien's displacement law states that the black body curve is shifted along the wavelength axis inversely in proportion to the value of T. To some this is apparent from staring at equation 1 for a few seconds. Those, like the author, less gifted at such matters must go through a more thorough analysis to convince themselves. Let us define the position of the Planck function along the wavelength axis by the value of λ at which the curve peaks, λ_{max}. The peak of the black body is defined by

$$\frac{\partial F}{\partial \lambda} = 0.$$

Differentiating equation 1, and ignoring ε, we have:

$$\frac{\partial F}{\partial \lambda} = 0 = \left\{ -\frac{4}{\lambda^5}\frac{1}{e^{c_2/\lambda T}-1} + \frac{1}{\lambda^6}\frac{c_2}{T}\frac{e^{c_2/\lambda T}}{(e^{c_2/\lambda T}-1)^2} \right\}_{\lambda = \lambda_{max}}$$

from which   6

$$1 - e^{-c_2/\lambda_{max}T} = \frac{c_2}{4\lambda_{max}\,T}$$

which we can solve to give an unique value of $\lambda_{max}T$. Hence $\lambda_{max}T$

is a constant equal to 3670 μm K, or $\lambda_{max} = 3670/T$.

This value differs from that in most physics books because we have used F instead of F_λ. The numerical factor 4 in the right hand side of equation 6 originates in the $1/\lambda^4$ term of equation 1. If we use equation 2 for F_λ, this becomes a factor 5, and λ_{max} is then 2898 μm K, the value given in most physics texts. By working in $F = \lambda F_\lambda$, we weight the position of the maximum of the Planck curve to longer wavelength.

APPENDIX II

UPDATE

This appendix contains a brief description of the more important developments in infrared astronomy between June 1974, when the main text of this book was finalised, and the end of that year. It would be an exaggeration to say that any startling breakthroughs occurred, but a number of the papers published in this period contain interesting results which have direct bearing on earlier work and which certainly merit description.

In July approval was given for Britain to build a 3·8 metre (150-inch) infrared telescope on Hawaii. This instrument, already dubbed UKIRT by the Americans, will be the largest dedicated infrared facility in the world. The cost will probably be about £1½ million, something like one-quarter the cost of a comparable optical telescope. The savings are mostly in the mirror, which will be figured to give images 2″ arc across instead of the ½″ arc required for optical use. Not only is the mirror figuring cheaper, but a thinner than normal blank can be used, and the reduction in weight considerably lowers the cost of the mechanical portions of the telescope. The UK infrared telescope will be sited near the summit of Mauna Kea at an altitude a little under 14,000 feet. It is expected that an American infrared telescope of smaller aperture will also be constructed on Mauna Kea.

A preliminary reduction of the Mariner 10 infrared observations of Mercury was published (Chase *et al., Science* 185, 142). The lowest temperature recorded on the dark side of Mercury was about 100 °K, in good agreement with the ground-based data described in chapter 4. This finally confirms that Mercury, like the moon, is dust covered. A few hot spots were also recorded, and by analogy with the moon we might assume that these are caused by boulders in and around the younger craters. This is perhaps a surprising result, however, for the extreme heating and cooling of Mercury's surface would be expected to cause fragmentation of boulders quite rapidly.

The argument that limb brightening of Jupiter at certain wavelengths implies the existence of a temperature inversion in the atmosphere was dispelled by Trafton and Wildey (*Astrophys. J.* 194, 499). These authors demonstrated that a layered atmosphere in which the temperature increases steadily inwards can provide an equally good fit to the observations. The same probably applies to other solar system bodies for which a temperature inversion has been implied.

New infrared photometry of Saturn's satellites was published by David Morrison (*Icarus* 22, 51). He deduced the diameters of Rhea and Dione to be 800 and 575 km respectively. This leads to a density of around $1\frac{1}{2}$ gm cm^{-3} for each, a figure sufficiently low that a considerable amount of ice must be present in both bodies. Morrison also examined Saturn's ring B (*ibid,* 57) and in particular the temperature shortly after the ring particles emerge from eclipse by the planet. For small particles this temperature drop depends on the size rather than the thermal inertia; from these observations Morrison deduced a minimum size for the ring particles of about 2 cm.

Uranus and Neptune have also been subjected to greater scrutiny, by Rieke and Low (*Astrophys. J.* 193, L147), who made use of the newly exploited 35 μm window. The brightness of the planets at these wavelengths is not well explained by current models, and this might eventually lead to the identification of other gases in the atmospheres of Uranus, Neptune and other bodies.

The migration of infrared astronomical research away from the cool stellar sources which were so popular at first is strikingly demonstrated by the lack of recent publications relating to Chapter 5, the only paper of note being Cohen and Fawley's failure to detect any further globular clusters at 10μm (*Mon. Not. Roy. Astron. Soc.* 169, 31P), which makes the detection of Messier 15 seem even more bizarre. Perhaps I may also mention a paper by Day (*Astrophys. J.* 192, L15) which has some relevance to cool stars. In seeking identifications for the "silicate bumps" and absorption features, many authors have assumed a crystalline structure to the grains and have consequently had to resort to complicated mixtures of complex minerals. Day showed that an excellent fit to the observed absorption data can be made using amorphous (non-crystalline) grains of magnesium silicate. These grains develop crystalline from (olivine, in fact) above 500 °K, which might impose some rather tight constraints on models of grain formation around stars.

Paula Szkody was the only person to tackle a new class of objects during the latter half of 1974. Observing from Kitt Peak she studied several of the dwarf novae and published results for SS Cygni and RX Andromedae (*Astrophys. J.* 192, L75). The dwarf novae are stars with rather distinctive light curves: for most of the time they remain fairly constant, but intermittently they brighten by a couple of magnitudes. Dwarf novae are explained by three interacting components: a white dwarf, a cool main sequence star and a gaseous ring. Surprisingly it is not the cool star which dominates the infrared emission, but free-free radiation from the gaseous disc. From the infrared observations Szkody was able to deduce a very high density for this disc.

Strom, Grasdalen and Strom finally published a fairly lengthy compendium of infrared and optical observations of Herbig-Haro objects (*Astrophys. J.* 191, 111) showing that they can be well explained by the reflections nebula hypothesis in which the nebula reflects the infrared source, a star hidden from us by intervening dust.

Another peculiar nebula in Cygnus, nicknamed "The Egg", was described by Ney in *Sky and Telescope* (49, 21). "The Egg" very closely resembles the Footprint Nebula, Minkowski 1–92 described in chapter 7. It is almost certainly of similar morphology – two bright lobes being illuminated through the poles of a flat disc so that when seen by us from earth the illuminating star is hidden by the dust in the disc. Unlike M1-92, the Egg does not have an emission-line spectrum; this suggests the star is rather cooler. It also has a more pronounced infrared excess, reaching −6 magnitude at 20 μm. Although within our galaxy, the Egg was mistaken for a pair of compact galaxies by Zwicky who catalogued it as IV Zw 2100 +36. Like the Footprint, the Egg is bright enough to be seen in many amateur telescopes.

The region of the strong radio and infrared H II region G333·6 −0·2 was photographed by Churms *et al* (*Mon. Not. Roy. Astron. Soc.* 169, 39P). A small nebula at the position of the infrared source was recorded; this nebula is highly reddened and obscured by intervening dust.

There has been a plethora of observations of the Orion Nebula in the 50 – 500 μm region. Of these the most interesting was by Soifer and Hudson (*Astrophys. J.* 191, L83) in which they examined the rather high flux around 400 μm and concluded that either (i) Messier 42 is more massive than had been thought, (ii) the Orion silicates radiate better at 400 μm than lunar dust or (iii) there's dust besides silicates in Orion. In practice the third alternative doesn't help much since no likely grain material is known which radiates any better than silicates at this wavelength.

REFERENCES

CHAPTER 1

1 Herschel, W. 'Investigation of the Powers of the Prismatic Colours to Heat and Illuminate Objects', *Phil. Trans. Roy. Soc. London* 90, 255–83 (1800).
   Herschel, W. 'Experiments on the Refrangibility of the Invisible Rays of the Sun', *Phil. Trans. Roy. Soc. London* 90, 284–92 (1800).
   Herschel, W. 'Experiments on the Solar, and on the Terrestrial Rays that Occasion Heat' part I, *Phil. Trans. Roy. Soc. London* 90, 293-326 (1800).
   part II, ibid, 437–538 (1800).

2 Smyth, C. P. *Teneriffe, an Astronomer's Experiment,* Lovell Reeve, London (1858).

3 Rosse, 4th Earl. 'On the Radiation of Heat from the Moon', *Proc, Roy. Soc. London* 17, 436–43 (1869).
   ibid, no II, *Proc. Roy. Soc. London* 19, 9–14 (1870).
   ibid, *Phil. Trans. Roy. Soc. London* 163, 587–627 (1873).

4 Coblentz, W. W. 'New Measurements of Stellar Radiation', *Astrophys. J.* 55, 20–3 (1922).
   Menzel, D. H., Coblentz, W. W. and Lampland, D. O. 'Planetary Temperatures Derived from Water-cell Transmissions', *Astrophys. J.* 63, 177–87 (1926).
   Pettit, E. and Nicholson, S. B. 'Stellar Radiation Measurements', *Astrophys. J.* 68, 279–308 (1928).
   Pettit, E. and Nicholson, S. B. 'Measurements of the Radiation from Variable Stars', *Astrophys. J.* 78, 320–53 (1933).
   Pettit, E. and Nicholson, S. B. 'Radiation from the Planet Mercury', *Astrophys. J.* 83, 84–102 (1936).

5 Pettit, E. and Nicholson, S. B. 'Lunar Radiation and Temperatures', *Astrophys. J.* 71, 102–35 (1930).
   Pettit, E. 'Lunar Radiation as Related to Phase', *Astrophys. J.* 81, 17–36 (1935).
   Pettit, E. 'Radiation Measurements on the Eclipsed Moon', *Astrophys. J.* 91, 408–20 (1940).

6 Wesselink, A. J. 'Heat Conductivity and Nature of the Lunar Surface Material', *Bull. Astron. Inst. Netherlands* 10, 351–63 (1948).

7 Golay, M. J. E. 'A Pneumatic Infrared Detector', *Rev. Sci. Instruments* 18, 357–62 (1947).

CHAPTER 2

1 Low, F. J. 'Low-temperature Germanium Bolometer', *J. Opt. Soc. America* 51, 1300–4 (1961).
2 Adel, A. 'The Extension of the Prismatic Solar Spectrum from 14µ to 24µ', *Astrophys. J.* 96, 239–41 (1942).
3 Becklin, E. E., Hansen, O., Kieffer, H. and Neugebauer, G. 'Stellar Flux Calibration at 10 and 20µ Using Mariner 6, 7 and 9 Results', *Astron. J.* 78, 1063–6 (1973).

CHAPTER 3

1 Felgett, P. B. 'An Exploration of Infra-red Stellar Magnitudes Using the Photoconductivity of Lead Sulphide', *Mon. Not. Roy. Astron. Soc.* 111, 537–59 (1951).
2 Johnson, H. L., Mitchell, R. I., Iriarte, B. and Wisniewski, W. Z. 'UBVRIJKL Photometry of the Bright Stars', *Comm. Lunar Planetary Lab.* 4, 99–110 (1966).
3 Johnson, H. L. 'Interstellar Extinction in the Galaxy', *Astrophys. J.* 141, 923–42 (1965).
4 Neugebauer, G. and Leighton, R. B. *Two-micron Sky Survey,* NASA SP-3047 (1969).
5 Neugebauer, G., Martz, D. E. and Leighton, R. B. 'Observations of Extremely Cool Stars', *Astrophys. J.* 142, 399–401 (1965).
6 Ulrich, B. T., Neugebauer, G., McCammon, D., Leighton, R. B., Hughes, E. E. and Becklin, E. E. 'Further Observations of Extremely Cool Stars', *Astrophys. J.* 146, 288–90 (1966).
7 Price, S. D. 'Results of an Infrared Stellar Survey', *Astron. J.* 73, 431–41 (1968).
8 Wildey, R. L. and Murray, B. C. '10-µ Photometry of 25 Stars from B8 to M7', *Astrophys. J.* 139, 435–41 (1964).
9 Ney, E. P. and Gould, R. J. 'Infrared Radiation from B stars', *Astrophys. J.* 140, 388–90 (1964). This result was refuted by Low, F. J. 'The Infrared Brightness of α Leonis and γ Orionis', *Astrophys. J.* 141, 326–7 (1965).
10 Mitchell, R. I. 'Nine-colour Photometry of ε Aur 0·35–9·5 µ', *Astrophys. J.* 140, 1607–8 (1964). Corrected by Low, F. J. and Mitchell, R. I. 'New Infrared Photometry of ε Aurigae', *Astrophys. J.* 141, 327–8 (1965).
11 Low, F. J., Rieke, G. H. and Armstrong, K. R. 'Ground-based Observations at 34 microns', *Astrophys. J.* 183, L105–L109 (1973).
12 Rieke, G. H., Harper, D. A., Low, F. J. and Armstrong, K. R. '350-micron Observations of Sources in H II Regions, the Galactic Center and NGC 253', *Astrophys. J.* 183, L67–L71 (1973).
13 Hoffmann, W. F., Frederick, C. L. and Emery, R. J. '100-micron Survey

of the Galactic Plane', *Astrophys. J.* 170, L89–L97 (1971).
14 The USAF survey was circulated privately to a few infrared astronomers. A much extended version is in preparation, but the date of its release has not been announced.

CHAPTER 4

1 Pettit, E. and Nicholson, S. B. 'Lunar Radiation and Temperatures', *Astrophys. J.* 71, 102–35 (1930).

2 Winter, D. F. and Saari, J. M. 'A Particulate Thermophysical Model of the Lunar Soil', *Astrophys. J.* 156, 1135–51 (1969).

3 Shorthill, R. W., Borough, H. C. and Conley, J. M. 'Enhanced Lunar Thermal Radiation During a Lunar Eclipse', *Publ. Astron. Soc. Pacific* 72, 481–5 (1960).

4 Saari, J. M. and Shorthill, R. W. 'Isotherms of Crater Regions on the Illuminated and Eclipsed Moon', *Icarus* 2, 115–36 (1963).

5 Saari, J. M., Shorthill, R. W. and Deaton, T. K. 'Infrared and Visible Images of the Eclipsed Moon of December 19, 1964', *Icarus* 5, 635–59 (1966).

6 Allen, D. A. 'Infrared Studies of the Lunar Terrain', part I *The Moon* 2, 320–37 (1971).
Part II ibid, 2, 435–62 (1971).

7 Allen, D. A. and Ney, E. P. 'Lunar Thermal Anomalies: Infrared Observations', *Science* 164, 419–20 (1969).

8 Murdock, T. L. and Ney, E. P. 'Mercury: the Dark-side Temperature', *Science* 170, 535–7 (1970).

9 Sinton, W. M. and Strong, J. 'Radiometric Observations of Mars', *Astrophys. J.* 131, 459–69 (1960).

10 Gatley, I., Kieffer, H.H, Miner, E. and Neugebauer, G. 'Infrared Observations of Phobos from Mariner 9', *Astrophys. J.* 190, 497–503 (1974).

11 Gillett, F. C., Low, F. J. and Stein, W. A. 'The 2·8–14-micron Spectrum of Jupiter', *Astrophys. J.* 157, 925–34 (1969).

12 Keay, C. S. L., Low, F. J., Rieke, G. H. and Minton, R. B. 'High-resolution Maps of Jupiter at 5 Microns', *Astrophys. J.* 183, 1063–73 (1973).

13 Westphal, J. A. 'Observations of Localised 5-micron Radiation from Jupiter', *Astrophys. J.* 157, L63–L64 (1969).
Westphal, J. A., Matthews, K. and Terrile, R. J. 'Five-micron Pictures of Jupiter', *Astrophys. J.* 188, L111–L112 (1974).

14 Gillett, F. C. and Westphal, J. A. 'Observations of 7·9-micron Limb Brightening on Jupiter', *Astrophys. J.* 179, L153–L154 (1974).

15 Murray, B. C., Wildey, R. L. and Westphal, J. A. 'Observations of Jupiter and the Galilean Satellites at 10 Microns', *Astrophys. J.* 139, 986–93 (1964).

16 Gillett, F. C., Merrill, K. M. and Stein, W. A. 'Albedo and Thermal Emission of Jovian Satellites I–IV', *Astrophys. Lett.* 6, 247–9 (1970).

17 Morrison, D. and Cruikshank, D. P. 'Thermal Properties of the

Galilean Satellites', *Icarus* 18, 224–36 (1973).

Hansen, O. 'Ten-micron Eclipse Observations of Io, Europa and Ganymede', *Icarus* 18, 237–46 (1973).

18 Fink, U., Dekkers, N. H. and Larson, H. P. 'Infrared Spectra of the Galilean Satellites of Jupiter', *Astrophys. J.* 179, L155–L159 (1973).

19 Gillett, F. C. and Forrest W. J. 'The 7·5- to 13·5-micron Spectrum of Saturn', *Astrophys. J.* 187, L37–L39 (1974).

20 Nolt, I. G., Radostitz, J. V., Donnelly, R. J., Murphy, R. E. and Ford, H. C. 'Thermal Emission of Saturn's Rings and Disk at 34 μm', *Nature* 248, 659–60 (1974).

21 Kuiper, G. P., Cruikshank, D. P. and Fink, U. *Sky and Telescope* 39, 80 (1970).

Pilcher, C. B., Chapman, C. R., Lebofsky, L. A. and Kieffer, H. H. 'Saturn's Rings: Identification of Water Frost', *Science* 167, 1372–3 (1970).

22 Allen, D. A. and Murdock, T. L. 'Infrared Photometry of Saturn, Titan and the Rings', *Icarus* 14, 1–2 (1971).

23 Gillett, F. C., Forrest, W. J. and Merrill, K. M. '8–13 Micron Observations of Titan', *Astrophys. J.* 184, L93–L95 (1973).

24 Low, F. J. 'The Infrared Brightness Temperature of Uranus', *Astrophys. J.* 146, 326–7 (1966).

25 Morrison, D. and Cruikshank, D. P. 'Temperatures of Uranus and Neptune at 24 Microns', *Publ. Astron. Soc. Pacific* 84, 642 (1972).

26 Ney, E. P. and Maas, R. W. 'Infrared Observations of Mercury and Uranus', *Bull. American Astron. Soc.* 1, 202 (1969).

27 Allen, D. A. 'Infrared Diameter of Vesta', *Nature* 227, 158–9 (1971).

28 Cruikshank, D. P. and Morrison, D. 'Radii and Albedos of Asteroids 1, 2, 3, 4, 6, 15, 51, 433 and 511', *Icarus* 20, 477–81 (1973).

Morrison, D. 'Radiometric Diameters and Albedos of 40 Asteroids', *Astrophys. J.* 194, 203–12 (1974).

29 Becklin, E. E. and Westphal, J. A. 'Infrared Observations of Comet 1965f', *Astrophys. J.* 145, 445–53 (1966).

30 Maas, R. W., Ney, E. P. and Woolf, N. J. 'The 10-micron Emission Peak of Comet Bennett 1969i', *Astrophys. J.* 160, L101–L104 (1970).

31 Ney, E. P. 'Infrared Observations of Comet Kohoutek near Perihelion', *Astrophys. J.* 189, L141–L143 (1974).

32 Rieke, G. H. and Lee, T. A. 'Photometry of Comet Kohoutek (1973f)', *Nature* 248, 737–40 (1974).

33 Rawcliffe, R. D., Bartky, C. D., Li, F., Gordon, E. and Carta, D. 'Meteor of August 10, 1972', *Nature* 247, 449–50 (1974).

34 Soifer, B. T., Houck, J. R. and Harwit, M. O. 'Rocket-infrared Observations of the Interplanetary Medium', *Astrophys. J.* 168, L73–L78 (1971).

CHAPTER 5

1 Johnson, H. L. 'Infrared Stars', *Sky and Telsecope* 32, 73–7 (1966); *see*

also Wing, R. F., Spinrad, H. and Kuhi, L. V. 'Spectrophotometric Studies of Three Extremely Red Stars', *Astron. J.* 71, 187 (1966).

2 Wisniewski, W. Z., Wing, R. F., Spinrad, H. and Johnson, H. L. 'Additional Observations of "Infrared Stars" ', *Astropyhs. J.* 148, L29–L32 (1967).

3 Hyland, A. R., Becklin, E. E., Neugebauer, G. and Wallerstein, G. 'Observations of the Infrared Object, VY Canis Majoris', *Astrophys. J.* 158, 619–28 (1969).

4 Johnson, H. L. 'The Colors of M supergiants', *Astrophys. J.* 149, 345–52 (1967).

5 Neugebauer, G., Sargent, W. L. W., Westphal, J. A. and Porter, F. C. '1·6μ–10μ Observations of R Doradus and W Hydrae', *Publ. Astron. Soc. Pacific* 83, 346–50 (1971).

6 Woolf, N. J. and Ney, E. P. 'Circumstellar Infrared Emission from Cool Stars', *Astrophys. J.* 155, L181–L184 (1969).

7 Low, F. J. and Krishna Swamy, K. S. 'Narrow-band Infrared Photometry of α Ori', *Nature* 227, 1333–4 (1970).

8 Gehrz, R. D., Ney, E. P. and Strecker, D. W. 'Observations of Anomalous Radiation at Long Wavelengths from Ic Class Variables', *Astrophys. J.* 161, L219–L223 (1970).

9 Gehrz, R. D. and Woolf, N. J. 'Mass Loss from M Stars', *Astrophys. J.* 165, 285–94 (1971).

10 Gillett, F. C., Stein, W. A. and Solomon, P. M. 'The Spectrum of VY Canis Majoris from 2·9 to 14 Microns', *Astrophys. J.* 160, L173–L176 (1970).

11 Gammon, R. H., Gaustad, J. E. and Treffers, R. R. 'Ten-micron Spectroscopy of Circumstellar Shells', *Astrophys. J.* 175, 687–91 (1972).

12 Humphreys, R. M., Strecker, D. W. and Ney, E. P. 'High Luminosity G Supergiants', *Astrophys. J.* 167, L35–L39 (1971).

13 Humphreys, R. M. and Ney, E. P. 'Infrared Stars in Binary Systems', *Astrophys. J.* 190, 339–47 (1974).

14 Gillett, F. C., Hyland, A. R. and Stein, W. A. '89 Herculis: an F2 Supergiant with Large Circumstellar Infrared Emission', *Astrophys. J.* 162, L21–L24 (1970).

15 Allen, D. A. and Ney, E. P. 'Is 17 Leporis a Shell Star?', *Observatory* 92, 47–9 (1972).

16 Gilman, R. C. 'On the Composition of Circumstellar Grains', *Astrophys. J.* 155, L185–L187 (1969).

17 Hyland, A. R., Becklin, E. E., Frogel, J. A. and Neugebauer, G. 'Infrared Observations of 1612 MHz IR/OH Sources', *Astron. Astrophys.* 16, 204–19 (1972).

18 Strecker, D. W., Ney, E. P. and Murdock, T. L. 'Cygnids and Taurids – Two Classes of Infrared Objects', *Astrophys. J.* 183, L13–L16 (1973).

19 Gilman, R. C. 'Free-free and Free-bound Emission in Low-surface-gravity Stars', *Astrophys. J.* 188, 87–94 (1974).

20 Frogel, J. A., Hyland, A. R., Kristian, J. and Neugebauer, G. 'The

Unusual Infrared Object IRC+10216', *Astrophys. J.* 158, L133–L137 (1969).

21 Miller, J. S. 'Scanner Observations of the Leo Infrared Object IRC+ 10216', *Astrophys. J.* 161, L95–L99 (1970).

22 Toombs, R. I., Becklin, E. E., Frogel, J. A., Law, S. K., Porter, F. C. and Westphal, J. A. 'Infrared Diameter of IRC+10216 Determined from Lunar Occultations', *Astrophys. J.* 173, L71–L74 (1972).

23 Gehrz, R. D. and Woolf, N. J. 'RV Tauri Stars: a New Class of Infrared Object', *Astrophys. J.* 161, L213–L217 (1970).
Gehrz, R. D. 'Infrared Radiation from RV Tauri Stars', *Astrophys. J.* 178, 715–25 (1972).

24 Stein, W. A., Gaustad, J. E., Gillett, F. C. and Knacke, R. F. 'Circumstellar Infrared Emission from Two Peculiar Objects – R Aquarii and R Coronae Borealis', *Astrophys. J.* 155, L3–L7 (1969).

25 Lee, T. A. and Feast, M. W. 'Infrared Excess of RY Sgr', *Astrophys. J.* 157, L173–L176 (1969).

26 Feast, M. W. and Glass, I. S. 'Infra-red Photometry of R Coronae Borealis Type Variables and Related Objects', *Mon. Not. Roy. Astron. Soc.* 161, 293–303 (1973).

27 Lee, T. A. and Nariai, K. 'Infrared Radiation from Upsilon Sagittarii', *Astrophys. J.* 149, L93–L95 (1967).

28 Forrest, W. J., Gillett, F. C. and Stein, W. A. 'Infrared Measurements of R Coronae Borealis through its 1972 March–June Minimum', *Astrophys. J.* 178, L129–L132 (1972).

29 Lee, T. A. 'Visual and Infrared Photometry of RY Sagittarii near the Phase of Deep Minimum', *Publ. Astron. Soc. Pacific* 85, 637–40 (1973).

30 Frogel, J. A., Kleinmann, D. E., Kunkel, W., Ney, E. P. and Strecker, D. W. 'Multicolor Photometry of the M Dwarf Proxima Centauri', *Publ. Astron. Soc. Pacific* 84, 581–2 (1972).   —

31 Lee, T. A. 'On the Nature of Some Faint Infrared Stars', *Astron. J.* 77, 374–5 (1972).

32 Woolf, N. J. 'Infrared Emission from Unusual Binary Stars', *Astrophys. J.* 185, 229–37 (1973).

33 Huang, S-S. 'An Interpretation of ε Aurigae', *Astrophys. J.* 141, 976–84 (1965).

34 Swings, J. P. and Allen, D. A. 'Photometry of Symbiotic and VV Cephei Stars in the Near Infrared', *Publ. Astron. Soc. Pacific* 84, 523–7 (1972).

35 Glass, I. S. and Webster, B. L. 'Infrared Photometry of RR Telescopii and Other Emission-line Objects', *Mon. Not. Roy. Astron. Soc.* 165, 77–89 (1973).

36 Webster, B. L. and Allen, D. A. 'Symbiotic Stars and Dust', *Mon. Not. Roy. Astron. Soc.* 171, 171–80 (1975).

37 Harvey, P. M. 'Infrared Variability of V1016 Cygni', *Astrophys. J.* 188, 95–6 (1974).

38 Allen, D. A. 'Infrared Observations of Northern Emission-line Stars',

Mon. Not. Roy. Astron. Soc. 168, 1–13 (1974).
39 Wray, J. D. 'Highly Reddened Objects in Sagittarius', *Astrophys. J.* 168, L97–L99 (1971).
40 MacGregor, A. D., Phillips, J. P. and Selby, M. J. 'The Detection of M 15 at 10·2 μ', *Mon. Not. Roy. Astron. Soc.* 164, 31P–33P (1973).

CHAPTER 6
1 Woolf, N. J., Stein, W. A. and Strittmatter, P. A. 'Infrared Emission from Be Stars', *Astron. Astrophys.* 9, 252–8 (1970).
2 Allen, D. A. 'Near Infra-red Magnitudes of 248 Early-type Emission-line Stars and Related Objects', *Mon. Not. Roy. Astron. Soc.* 161, 145–66 (1973).
3 There have been many papers describing planetary nebulae in the PbS region. The first of these were:
Khromov, G. S. and Moroz, V. I. 'Infrared Radiation of Planetary Nebulae', *Soviet Astron. – A.J.* 15, 892–900 (1972. Russian original dated 1971) and Willner, S. P., Becklin, E. E. and Visvanathan, N. 'Observations of Planetary Nebulae at 1·65 to 3·4 Microns', *Astrophys. J.* 175, 699–706 (1972).
4 Low, F. J., Johnson, H. L., Kleinmann, D. E., Latham, A. S. and Geisel, S. L. 'Photometric and Spectroscopic Observations of Infrared Stars', *Astrophys. J.* 160, 531–43 (1970).
5 Geisel, S. L. 'Infrared Excesses, Low-excitation Emission Lines, and Mass Loss', *Astrophys. J.* 161, L105–L108 (1970).
6 Swings, J. P. and Allen, D. A. 'The Infrared Object HD 45677', *Astrophys. J.* 167, L41–L45 (1971).
7 Allen, D. A. and Swings, J. P. 'Infrared Excesses and Forbidden Emission Lines in Early-type Stars', *Astrophys. Lett.* 10, 83–7 (1972).
8 Neugebauer, G. and Westphal, J. A. 'Infrared Observations of Eta Carinae', *Astrophys. J.* 152, L89–L94 (1968).
9 Swings, J. P. and Allen, D. A. 'MWC 645 and MWC 819: Two Stars Resembling Eta Carinae', *Astrophys. Lett.* 14, 65–8 (1973).
10 Allen, D. A. 'Infrared Observations of Northern Emission-line Stars', *Mon. Not. Roy. Astron. Soc.* 168, 1–13 (1974).
11 Robinson, G., Hyland, A. R. and Thomas, J. A. 'Observations and Interpretation of the Infrared Spectrum of Eta Carinae', *Mon. Not. Roy. Astron. Soc.* 161, 281–92 (1973).
Gehrz, R. D., Ney, E. P., Becklin, E. E. and Neugebauer, G. 'The Infrared Spectrum and Angular Size of Eta Carinae', *Astrophys. Lett.* 13, 89–93 (1973).
12 Humphreys, R. M., Strecker, D. W., Murdock, T. L. and Low, F. J. 'IRC +10420 – Another Eta Carinae?', *Astrophys. J.* 179, L49–L52 (1973).
13 Dyck, H. M. and Milkey, R. W. 'Infrared Excesses in Early-type Stars: Free-free Emission', *Publ. Astron. Soc. Pacific* 84, 597–612 (1972).

14 Milkey, R. W. and Dyck, H. M. 'Low-temperature Free-free Emission: Infrared Excesses in Be Stars', *Astrophys. J.* 181, 833–9 (1973).

15 Humphreys, R. M. and Ney, E. P. 'Infrared Stars in Binary Systems', *Astrophys. J.* 190, 339–47 (1974).

16 Allen, D. A. and Penston, M. V. 'Dust Temperatures around Hot Stars', *Mon. Not. Roy. Astron. Soc.* 165, 121–31 (1973).

17 Gillett, F. C. and Merrill, K. M. 'Infrared Studies of Galactic Nebulae', *Astrophys. J.* 172, 367–74 (1972).

18 Cohen, M. and Barlow, M. J. 'An Infrared Photometric Survey of Planetary Nebulae', *Astrophys. J.* 193, 401–8 (1974).

19 Allen, D. A. and Glass, I. S. 'Infrared Photometry of Southern Emission-line Stars', *Mon. Not. Roy. Astron. Soc.* 167, 337–50 (1974).

20 Allen, D. A. and Swings, J. P. 'The Peculiar Nebula M2-9', *Astrophys. J.* 174, 583–9 (1972).

21 Glass, I. S. and Webster, B. L. 'Infra-red Photometry of RR Telescopii and Other Emission-line Objects', *Mon. Not. Roy. Astron. Soc.* 165, 77–89 (1973).

22 Allen, D. A., Harvey, P. M. and Swings, J. P. 'Infrared Photometry of Northern Wolf-Rayet Stars', *Astron. Astrophys.* 20, 333–6 (1972).
    Allen, D. A. and Porter, F. C. 'Infrared Photometry of Southern Wolf-Rayet Stars', *Astron. Astrophys.* 22, 159–60 (1973).

23 Webster, B. L. and Glass, I. S. 'The Coolest Wolf-Rayet Stars', *Mon. Not. Roy. Astron. Soc.* 166, 491–7 (1974).

24 Hyland, A. R. and Neugebauer, G. 'Infrared Observations of Nova Serpentis 1970', *Astrophys. J.* 160, L177–L180 (1970).
    Geisel, S. L., Kleinmann, D. E. and Low, F. J. 'Infrared Emission of Novae', *Astrophys. J.* 161, L101–L104 (1970).

25 Kirshner, R. P., Willner, S. P., Becklin, E. E., Neugebauer, G. and Oke, J. B. 'Spectrophotometry of the Supernova in NGC 5253 from 0·33 to 2·2 Microns', *Astropys. J.* 180, L97–L100 (1973).

26 Ney, E. P. and Stein, W. A. 'Observations of the Crab Nebula at $\lambda =$ 5800 Å, 2·2 μ and 3·5 μ with a 4-minute Beam', *Astrophys. J.* 152, L21–L24 (1968).

27 Becklin, E. E. and Kleinmann, D. E. 'Infrared Observations of the Crab Nebula', *Astrophys. J.* 152, L25–L30 (1968).

28 Aitken, D. K. and Polden, P. G. 'Measurement of the 10 μm Flux from the Crab Nebula', *Nature Phys. Sci.* 233, 45–6 (1971).
    Aitken, D. K. and Polden, P. G. 'Dust in the Crab Nebula', *Nature Phys. Sci.* 234, 72–3 (1971).

29 Becklin, E. E., Kristian, J., Matthews, K. and Neugebauer, G. 'Measurement of the Crab Pulsar at 2·2 and 3·5 Microns', *Astrphys. J.* 186, L137–L139 (1973).

30 Neugebauer, G., Oke, J. B., Becklin, E. E. and Garmire, G. 'A Study of Visual and Infrared Observations of Sco XR-1', *Astrophys. J.* 155, 1–9 (1969).

31 Penny, A. J., Olowin, R. P., Penfold, J. E. and Warren, P. R. 'Photo-

metry of HD 153919 = 2U 1700–37', *Mon. Not. Roy. Astron. Soc.* 163, 7P–12P (1973).

32 Frogel, J. A. and Persson, S. E. 'Infrared Photometry of the X-ray Sources 2U 0900–40 and 2U 1700–37', *Publ. Astron. Soc. Pacific* 85, 641–2 (1973).
33 Hyland, A. R. and Mould, J. R. 'Infrared Variability and the Interstellar Reddening of the X-ray Source HD 77581', *Astrophys. J.* 186, 993–6 (1973).
34 Becklin, E. E., Kristian, J., Neugebauer, G. and Wynn-Williams, C. G. 'Discovery of Infrared Emission from the Radio Source near Cygnus X–3', *Nature Phys. Sci.* 239, 130 (1972).
35 Becklin, E. E., Neugebauer, G., Hawkins, F. J., Mason, K. O., Sanford, P. W., Matthews, K. and Wynn-Williams, C. G. 'Infrared and X-ray Variability of Cyg X–3', *Nature* 245, 302–3 (1973).
36 Johnson, H. L. 'Interstellar Extinction in the Galaxy', *Astrophys. J.* 141, 923–42 (1965).

CHAPTER 7

1 Mendoza V., E. E. 'Infrared Photometry of T Tauri Stars and Related Objects', *Astrophys. J.* 143, 1010–14 (1966).
2 Mendoza V., E. E. 'Infrared Excesses in T Tauri Stars and Related Objects', *Astrophys. J.* 151, 977–89 (1968).
3 Strom, S. E., Strom, K. M., Brooke, A. L., Bregman, J. and Yost, J. 'Circumstellar Shells in the Young Cluster NGC 2264', *Astrophys. J.* 171, 267–78 (1972).
4 Knacke, R. F., Strom, S. E., Strom, K. M. and Young, E. 'Infrared Studies of the Young Clusters IC 2944, NGC 6530, NGC 6611 and Orion I', *Bull. American Astron. Soc.* 4, 325 (1972).
Knacke, R. F., Strom, K. M., Strom, S. E., Young, E. and Kunkel, W. 'A Young Stellar Group in the Vicinity of R Coronae Austrinae', *Astrophys. J.* 179, 847–54 (1973).
5 Penston, M. V. 'Multicolor Observations of Stars in the Vicinity of the Orion Nebula', *Astrophys. J.* 183, 505–34 (1973).
6 Ney, E. P., Strecker, D. W. and Gehrz, R. D. 'Dust Emission Nebulae around Orion O and B Stars', *Astrophys. J.* 180, 809–16 (1973).
7 Allen, D. A. and Penston, M. V. 'Dust Temperatures around Hot Stars', *Mon. Not. Roy. Astron. Soc.* 165, 121–31 (1973).
8 Cohen, M. 'Infrared Observations of Young Stars', *Mon. Not. Roy. Astron. Soc.* 161, 85–111 (1973). In three parts.
9 Glass, I. S. and Penston, M. V. 'An Infrared Survey of RW Aurigae Stars', *Mon. Not. Roy. Astron. Soc.* 167, 237–49 (1974).
10 Gillett, F. C. and Stein, W. A. 'Infrared Studies of Galactic Nebulae', *Astrophys. J.* 164, 77–82 (1971).
11 Strom, S. E., Strom, K. M., Yost, J., Carrasco, L. and Grasdalen, G. L. 'The Nature of the Herbig Ae- and Be-type Stars Associated with Nebulosity', *Astrophys. J.* 173, 353–66 (1972).

12 Allen, D. A. 'Near Infra-red Magnitudes of 248 Early-type Emission-line Stars and Related Objects', *Mon. Not. Roy. Astron. Soc.* 161, 145–66 (1973).

13 Cohen, M. and Dewhirst, D. W. 'Two Infrared Sources in Nebulae', *Nature* 228, 1077 (1970).

14 Cohen, M. and Woolf, N. J. 'Two Young Bright Infrared Objects', *Astrophys. J.* 169, 543–7 (1971).

15 Cohen, M. 'Optical Identifications of Infrared Sources', *Astrophys. Lett.* 9, 95–100 (1971).

16 Rieke, G. H., Lee, T. A. and Coyne, G. V. 'Photometry and Polarimetry of V1057 Cygni', *Publ. Astron. Soc. Pacific* 84, 37–45 (1972).
Simon, T., Morrison, N. D., Wolff, S. C. and Morrison, D. 'Far-infrared and uvby Photometry of V1057 Cygni', *Astron. Astrophys.* 20, 99–104 (1972).

17 Strom, S. E. 'Optical and Infrared Observations of Young Stellar Objects – an Informal Review', *Publ. Astron. Soc. Pacific* 84, 745–56 (1972).

18 Feast, M. W. and Glass, I. S. 'The Nature of a Nebulous Object in the Chamaeleon T Association', *Mon. Not. Roy. Astron. Soc.* 164, 35P–38P (1973).

19 Strom, K. M., Strom, S. E. and Grasdalen, G. L. 'An infrared source Associated with a Herbig–Haro Object', *Astrophys. J.* 187, 83–6 (1974).

20 Allen, D. A. and Swings, J. P. 'Infrared Excesses and Forbidden Emission Lines in Early-type Stars', *Astrophys. Lett.* 10, 83–7 (1972).

21 Glass, I. S. 'Observations of 30 Doradus in the Infrared', *Nature Phys. Sci.* 237, 7–8 (1972).

22 Low, F. J. and Aumann, H. H. 'Observations of Galactic and Extra-galactic Sources between 50 and 300 Microns', *Astrophys. J.* 162, L79–L85 (1970).

23 Harper, D. A. and Low, F. J. 'Far-infrared Emission from H II Regions', *Astrophys. J.* 165, L9–L13 (1971).

24 Hoffmann, W. F., Frederick, C. L. and Emery, R. J. '100-micron Survey of the Galactic Plane', *Astrophys. J.* 170, L89–L97 (1971).

25 Emerson, J. P., Jennings, R. E. and Moorwood, A. F. M. 'Far-infrared Observations of H II Regions from Balloon Altitudes', *Astrophys. J.* 184, 401–14 (1973).

26 Erickson, E. F., Swift, C. D., Witteborn, F. C., Mord, A. J., Auguson, G. C., Caroff, L. J., Kunz, L. W. and Giver, L. P. 'Infrared Spectrum of the Orion Nebula between 55 and 200 Microns', *Astrophys. J.* 183, 535–9 (1973).

27 Soifer, B. T., Pipher, J. L. and Houck, J. R. 'Rocket Infrared Observations of H II Regions', *Astrophys. J.* 177, 315–23 (1972).

28 Olthof, H. and van Duinen, R. J. 'Two Colour Far Infrared Photometry of Some Galactic H II Regions', *Astron. Astrophys.* 29, 315–20 (1973).

29 Kleinmann, D. E. and Low, F. J. 'Discovery of an Infrared Nebula in

Orion', *Astrophys. J.* 149, L1–L4 (1967).

30 Lemke, D. and Low, F. J. '21-micron Observations of H II Regions', *Astrophys. J.* 177, L53–L58 (1972).

Kleinmann, D. E. 'Bright Infrared Sources in M 17', *Astrophys. Lett.* 13, 49–54 (1973).

31 Ney, E. P. and Allen, D. A. 'The Infrared Sources in the Trapezium Region of M 42', *Astrophys. J.* 155, L193–L196 (1969).

32 Stein, W. A. and Gillett, F. C. 'Spectral Distribution of Infrared Radiation from the Trapezium Region of the Orion Nebula', *Astrophys. J.* 155, L197–L199 (1969).

33 Gillett, F. C. and Stein, W. A. 'Infrared Studies of Galactic Nebulae', *Astrophys. J.* 159, 817–22 (1970).

Woolf, N. J., Stein, W. A., Gillett, F. C., Merrill, K. M., Becklin, E. E., Neugebauer, G. and Pepin, T. J. 'The Infrared Sources in M 8', *Astrophys. J.* 179, L111–L115 (1973).

34 Wynn-Williams, C. G., Becklin, E. E. and Neugebauer, G. 'Infra-red Sources in the H II Region W3', *Mon. Not. Roy. Astron. Soc.* 160, 1-14 (1972).

35 Becklin, E. E., Neugebauer, G. and Wynn-Williams, C. G. 'Infrared Emission from the OH/H_2O Sources in W 49', *Astrophys. Lett.* 13, 147–9 (1973).

Wynn-Williams, C. G., Becklin, E. E. and Neugebauer, G. 'Infrared Studies of H II Regions and OH Sources', *Astrophys. J.* 187, 473–85 (1974).

Becklin, E. E., Frogel, J. A., Kleinmann, D. E., Neugebauer, G., Persson, S. E. and Wynn-Williams, C. G. 'Infrared Emission from the Southern H II Region H2–3', *Astrophys. J.* 187, 487–9 (1974).

36 Gillett, F. C., Low, F. J. and Stein, W. A. 'Infrared Observations of the Planetary Nebula NGC 7027', *Astrophys. J.* 149, L97–L100 (1967).

37 Knacke, R. F. and Dressler, A. M. 'The Spatial Distribution of the 11·7 Micron Radiation of NGC 7027', *Publ. Astron. Soc. Pacific* 85, 100–2 (1973).

Becklin, E. E., Neugebauer, G. and Wynn-Williams, C. G. 'The Spatial Distribution of Infrared Emission from NGC 7027', *Astrophys. Lett.* 15, 87–90 (1973).

38 Aitken, D. K. and Jones, B. 'Some Features of the Infra-red Spectrum of NGC 7027', *Mon. Not. Roy. Astron. Soc.* 165, 363–8 (1973).

39 Neugebauer, G. and Garmire, G. 'Infrared Observations of the Nebula K3–50', *Astrophys. J.* 161, L91–L94 (1970).

40 Frogel, J. A. and Persson, S. E. 'Studies of Small H II Regions', *Astrophys. J.* 178, 667–72 (1972).

Frogel, J. A. and Persson, S. E. 'Infrared Photometry of Small HII Regions', *Mém. Roy. Soc. Sci. Liège* 5, 341–7 (1973).

Frogel, J. A. and Persson, S. E. 'Compact Infrared Sources Associated with Southern H II Regions', *Astrophys. J.* 192, 351–68 (1974).

41 Danziger, I. J., Frogel, J. A. and Persson, S. E. 'Observations of NGC

6302 from 0·35 to 20 Microns', *Astrophys. J.* 184, L29–L32 (1973).

42 Becklin, E. E., Frogel, J. A., Neugebauer, G., Persson, S. E. and Wynn-Williams, C. G. 'The H II Region G333·6–0·2, a Very Powerful 1–20 Micron Source', *Astrophys. J.* 182, L125–L129 (1973).

CHAPTER 8

1 Johnson, H. L. 'Infrared Photometry of Galaxies', *Astrophys. J.* 143, 187–91 (1965).

2 Penston, M. V. 'The V–K Colours of the Nuclei of Bright Galaxies', *Mon. Not. Roy. Astron. Soc.* 162, 359–66 (1973).
Glass, I. S. 'The JHKL Colours of Galaxies', *Mon. Not. Roy. Astron. Soc.* 164, 155–68 (1973).

3 Sandage, A. R., Becklin, E. E. and Neugebauer, G. 'UBVRIHKL Photometry of the Central Region of M 31', *Astrophys. J.* 157, 55–68 (1969).

4 Glass, I. S. 'Infrared Observations of NGC 7552 and NGC 7582', *Mon. Not. Roy. Astron. Soc.* 162, 35P–37P (1973).

5 Becklin, E. E., Fomalont, E. B. and Neugebauer, G. 'Infrared and Radio Observations of the Nucleus of NGC 253', *Astrophys. J.* 181, L27–L31 (1973).

6 Kleinmann, D. E. and Low, F. J. 'Observations of Infrared Galaxies', *Astrophys. J.* 159, L165–L172 (1970).

7 Rieke, G. H. and Low, F. J. 'Infrared Photometry of Extragalactic Sources', *Astrophys. J.* 176, L95–L100 (1972).

8 Becklin, E. E., Frogel, J. A., Kleinmann, D. E., Neugebauer, G., Ney, E. P. and Strecker, D. W. 'Infrared Observations of the Core of Centaurus A, NGC 5128', *Astrophys. J.* 170, L15–L19 (1971).

9 Wisniewski, W. Z. and Kleinmann, D. E. 'Multicolor Photometry of Seyfert Galaxies and Measurement at 1·55μ of the Jet in M 87', *Astron. J.* 73, 866–7 (1968).

10 Oke, J. B., Sargent, W. L. W., Neugebauer, G. and Becklin, E. E. 'A Variable Radio-quiet Compact Galaxy I Zw 1727+50', *Astrophys. J.* 150, L173–L176 (1967).
Zwicky, F., Oke, J. B., Neugebauer, G., Sargent, W. L. W. and Fairall, A. P. 'The Variable Compact Galaxy Zw 0039·5+4003', *Publ. Astron. Soc. Pacific* 82, 93–8 (1970).

11 Neugebauer, G., Becklin, E. E. and Hyland, A. R. 'Infrared Sources of Radiation', *Ann. Rev. Astron. Astrophys.* 9, 67–102 (1971).

12 Young, E. T., Knacke, R. F. and Joyce, R. R. 'Infrared Photometry of Markarian 231', *Nature* 238, 263 (1972).

13 Johnson, H. L. 'The Brightness of 3C 273 at 2·2 μ', *Astrophys. J.* 139, 1022–3 (1964).

14 Low, F. J. and Johnson, H. L. 'The Spectrum of 3C 273', *Astrophys. J.* 141, 336–7 (1965).

15 Oke, J. B., Neugebauer, G. and Becklin, E. E. 'Absolute Spectral Energy Distribution of Quasi-stellar Objects from 0·3 to 2·2 Microns',

216

Astrophys. J. 159, 341–55 (1970).

16 Oke, J. B., Neugebauer, G. and Becklin, E. E. 'Spectrophotometry and Infrared Photometry of BL Lacertae', *Astrophys. J.* 156, L41–L43 (1969).

17 Stein, W. A., Gillett, F. C. and Knacke, R. F. 'Possible Upper Limit to the Distance of BL Lacertae', *Nature* 231, 254–5 (1971).

18 Dyck, H. M., Kinman, T. D., Lockwood, G. W. and Landolt, A. U. 'Observations of OJ 287 between 0·36 and 3·4 μm', *Nature Phys. Sci.* 234, 71–2 (1971).
Strittmatter, P. A., Serowski, K., Carswell, R. F., Stein, W. A., Merrill, K. M. and Burbridge, E. M. 'Compact Extagalactic Nonthermal Sources', *Astrophys. J.* 175, L7–L13 (1972).

19 Pacholczyk, A. G. and Wisniewski, W. Z. 'Infrared Radiation from the Seyfert Galaxy NGC 1068', *Astrophys. J.* 147, 394–6 (1967).

20 Low, F. J. and Kleinmann, D. E. 'Infrared Observations of Seyfert Galaxies, Quasistellar Sources and Planetary Nebulae', *Astron. J.* 73, 868–9 (1968).

21 Neugebauer, G., Garmire, G., Rieke, G. H. and Low, F. J. 'Infrared Observations on the Size of NGC 1068', *Astrophys. J.* 166, L45–L47 (1971).

22 Becklin, E. E., Matthews, K., Neugebauer, G. and Wynn-Williams, C. G. 'The Size of NGC 1068 at 10 Microns', *Astrophys. J.* 186, L69–L72 (1973).

23 Kleinmann, D. E. and Low, F. J. 'Infrared Observations of Galaxies and of the Extended Nucleus in M 82', *Astrophys. J.* 161, L203–L206 (1970).

24 Low, F. J. and Rieke, G. H. 'Variations in the 10 μm Flux from NGC 1068', *Nature* 233, 256–7 (1971).

25 Rieke, G. H. and Low, F. J. 'Variability of Extragalactic Sources at 10 Microns', *Astropys. J.* 177, L115–L119 (1972).

26 Stein, W. A., Gillett, F. C. and Merrill, K. M. 'Observations of the Infrared Radiation from the Nuclei of NGC 1068 and NGC 4151', *Astrophys. J.* 187, 213–7 (1974).

27 Fitch, W. S., Pacholczyk, A. G. and Weymann, R. J. 'Light Variations of the Seyfert Galaxy NGC 4151', *Astrophys. J.* 150, L67–L70 (1967).
Pacholczyk, A. G. 'Light Variations of the Seyfert Galaxy NGC 4151', *Astrophys. J.* 163, 449–54 (1971).

28 Penston, M. V., Penston, M. J., Neugebauer, G., Tritton, K. P., Becklin, E. E. and Visvanathan, N. 'Observations of NGC 4151 during 1970 in the Optical and Infra-red', *Mon. Not. Roy. Astron. Soc.* 153, 29–40 (1971).
Penston, M. V., Penston, M. J., Selmes, R. A., Becklin, E. E. and Neugebauer, G. 'Broad-band Optical and Infrared Observations of Seyfert Galaxies', *Mon. Not. Roy. Astron. Soc.* 169, 357–93 (1974).

29 Stein, W. A. and Gillett, F. C. 'Possible Variations of λ = 10 μm Radiation from NGC 4151', *Nature* 224, 675–6 (1969).

Stein, W. A. and Gillett, F. C. 'Photometric Measurements at $\lambda = 11$ μm of NGC 4151', *Nature Phys. Sci.* 233, 16–17 (1971).

30 Low, F. J. and Aumann, H. H. 'Observations of Galactic and Extragalactic Sources between 50 and 300 Microns', *Astrophys. J.* 162, L79–L85 (1970).

31 Low, F. J. 'The Infrared-galaxy Phenomenon', *Astrophys. J.* 159, L173–L177 (1970).

32 Harper, D. A. and Low, F. J. 'Far-infrared Observations of Galactic Nuclei', *Astrophys. J.* 182, L89–L93 (1973).

33 Jameson, R. F., Longmore, A. J., McLinn, J. A. and Woolf, N. J. 'Infrared Spectrum of NGC 1068', *Astrophys. J.* 187, L109 (1974).
Jameson, R. F., Longmore, A. J., McLinn, J. A. and Woolf, N. J. 'Infrared Emission by Dust in NGC 1068 and Three Planetary Nebulae' *Astrophys. J.* 190, 353–7 (1974).

34 Maffei, P. 'Infrared Object in the Region of IC 1805', *Publ. Astron. Soc. Pacific* 80, 618–21 (1968).

35 Spinrad, H., Sargent, W. L. W., Oke, J. B., Neugebauer, G., Landau, R., King, I. R., Gunn, J. E., Garmire, G. and Dieter, N. H. 'Maffei 1: a New Massive Member of the Local Group?', *Astrophys. J.* 163, L25–L31 (1971).

36 Spinrad, H., Bahcall, J., Becklin, E. E., Gunn, J. E., Kristian, J., Neugebauer, G., Sargent, W. L. W. and Smith, H. 'Optical and Near-infrared Observations of the nearby Spiral Galaxy Maffei 2', *Astrophys. J.* 180, 351–8 (1973).

37 Becklin, E. E. and Neugebauer, G. 'Infrared Observations of the Galactic Center', *Astrophys. J.* 151, 145–61 (1968).

38 Becklin, E. E. and Neugebauer, G. '1·65–19·5–micron Observations of the Galactic Center', *Astrophys. J.* 157, L31–L36 (1969).

39 Rieke, G. H. and Low, F. J. 'Map of the Galactic Nucleus at 10 μm', *Nature* 233, 53–4 (1971).

40 Hoffmann, W. F., and Frederick, C. L. 'Far-infrared Observation of the Galactic Center Region at 100 Microns', *Astrophys. J.* 155, L9–L13 (1969).
Hoffmann, W. F., Frederick, C. L. and Emery, R. J. '100-micron Map of the Galactic Center Region', *Astrophys. J.* 164, L23–L28 (1971).

41 Houck, J. R., Soifer, B. T., Pipher, J. L. and Harwit, M. O. 'Rocket-infrared Four-color Photometry of the Galaxy's Central Regions', *Astrophys. J.* 169, L31–L34 (1971).

42 Soifer, B. T. and Houck, J. R. 'Rocket-infrared Observations of the Galactic Center', *Astrophys. J.* 186, 169–76 (1973).

43 Friedlander, M. W. and Joseph, R. D. 'Detection of Celestial Sources at Far-infrared Wavelengths', *Astrophys. J.* 162, L87–L91 (1970).
Hoffman, W. F., Frederick, C. L. and Emery, R. J. '100-micron Survey of the Galactic Plane', *Astrophys. J.* 170, L89–L97 (1971).

44 Shivanandan, K., Houck, J. R. and Harwit, M. O. 'Preliminary Observations of the Far-infrared Night-sky Background Radiation',

Phys. Rev. Lett. 21, 1460–2 (1968).

Houck, J. R. and Harwit, M. O. 'Far-infrared Nightsky Emission above 120 Kilometers', *Astrophys. J.* 157, L45–L48 (1969).

45 Harwit, M. O., Houck, J. R. and Wagoner, R. V. 'Observational Upper Limits to the Electromagnetic Energy Radiated by External Galaxies', *Nature* 228, 451–2 (1970).

46 Pipher, J. L., Houck, J. R., Jones, B. W. and Harwit, M. O. 'Submillimetre Observations of the Night Sky Emission above 120 Kilometres', *Nature* 231, 375–8 (1971).

47 Muehlner, D. and Weiss, R. 'Measurement of the Isotropic Background Radiation in the Far Infrared', *Phys. Rev. Lett.* 24, 742–6 (1970).

CHAPTER 9

1 Allen, D. A. 'Infrared Objects in H II Regions', *Astrophys. J.* 172, L55–L58 (1972).

2 Allen, D. A. 'Optical Observations of Three New Infrared Sources', *Astrophys. Lett.* 12, 231–4 (1972).

3 Grasdalen, G. L., Strom, K. M. and Strom, S. E. 'A 2-micron Map of the Ophiuchus Dark-cloud Region', *Astrophys. J.* 184, L53–L57 (1973).

4 Simon, M., Righini, G., Joyce, R. R. and Gezari, D. Y. 'A Strong 350-micron Source in the Ophiuchus Dark Cloud', *Astrophys. J.* 186, L127–L130 (1974).

5 Ney, E. P. and Allen, D. A. 'The Infrared Sources in the Trapezium Region of M 42', *Astrophys. J.* 155, L193–L196 (1969).

6 Becklin, E. E. and Neugebauer, G. 'Observations of an Infrared Star in the Orion Nebula', *Astrophys. J.* 147, 799–802 (1967).

7 Kleinmann, D. E. and Low, F. J. 'Discovery of an Infrared Nebula in Orion', *Astrophys. J.* 149, L1–L4 (1967).

8 Rieke, G. H., Low, F. J. and Kleinmann, D. E. 'High-resolution Maps of the Kleinmann-Low Nebula in Orion', *Astrophys. J.* 186, L7–L11 (1973).

9 Wynn-Williams, C. G. and Becklin, E. E. 'Infrared Emission from H II Regions', *Publ. Astron. Soc. Pacific* 86, 5–25 (1974).

10 Gatley, I., Becklin, E. E., Matthews, K., Neugebauer, G., Penston, M. V. and Scoville, N. 'A New Infrared Complex and Molecular Cloud in Orion', *Astrophys. J.* 191, L121–L125 (1974).

11 Ney, E. P., Strecker, D. W. and Gehrz, R. D. 'Dust Emission Nebulae around Orion O and B Stars', *Astrophys. J.* 180, 809–16 (1973).

12 Penston, M. V., Allen, D. A. and Hyland, A. R. 'The Nature of Becklin's Star', *Astrophys. J.* 170, L33–L37 (1971).

13 Gillett, F. C. and Forrest, W. J. 'Spectra of the Becklin–Neugebauer Point Source and the Kleinmann–Low Nebula from 2·8 to 13·5 Microns', *Astrophys. J.* 179, 483–91 (1973).

14 Lemke, D., Low, F. J. and Thum, C. 'Infrared Map of the Orion

Nebula', *Astron. Astrophys.* 32, 231–33 (1974).

15 Becklin, E. E. Neugebauer, G. and Wynn-Williams, C. G. 'On the Nature of the Infrared Point Source in the Orion Nebula', *Astrophys. J.* 182, L7–L9 (1973).

16 Allen, D. A. and Penston, M. V. 'Infrared Stars or Infrareddened Stars?', *Nature* 251, 110–2 (1974).

17 Aitken, D. K. and Jones, B. 'Observations of the Infrared Extinction of IRS 5 in W3 Compared with the Galactic Center and the Becklin–Neugebauer Object', *Astrophys. J.* 184, 127–33 (1973).

18 Knacke, R. F., Cudaback, D. D. and Gaustad, J. E. 'Infrared Spectra of Highly Reddened Stars: a Search for Interstellar Ice Grains', *Astrophys. J.* 158, 151–60 (1969).

19 Stein, W. A. and Gillett, F. C. 'Search for Interstellar Silicate Absorption in Spectrum of VI Cyg No. 12', *Nature Phys. Sci.* 233, 72–3 (1971).

20 Hackwell, J. A., Gehrz, R. D. and Woolf, N. J. 'Interstellar Silicate Absorption Bands', *Nature* 227, 822–3 (1970).

21 Wynn-Williams, C. G., Becklin, E. E. and Neugebauer, G. 'Infrared Sources in the H II Region W3', *Mon. Not. Roy. Astron. Soc.* 160, 1–14 (1972).

22 Hoyle, F., Solomon, P. M. and Woolf, N. J., 'Massive Stars and Infrared Sources', *Astrophys. J.* 185, L89–L93 (1973).

23 The observations of ice and silicate absorption in DAA 6 were made by Gillett and Merrill, Aitken and Jones, and Allen. Publication is being delayed pending further measurements.

24 Schmidt, E. G. 'Optical Observations of an Infrared Source in NGC 2264', *Astrophys. J.* 176, L69–L71 (1972).

25 Merrill, K. M. and Soifer, B. T. 'Spectrophotometric Observations of a Highly Absorbed Object in Cygnus', *Astrophys. J.* 189, L27–L30 (1974).

26 Cohen, M. 'An Unusual Infrared Source near the Rosette Nebula', *Astrophys. J.* 185, L75–L78 (1973).

27 Borgman, J. 'The $9 \cdot 7\mu$ Absorption Feature in the Galactic Center', *Astron. Astrophys.* 29, 443–5 (1973).

28 Becklin, E. E., Neugebauer, G. and Wynn-Williams, C. G. 'Infrared Emission from the OH/H_2O Sources in W 49', *Astrophys. Lett.* 13, 147–9 (1973).

29 Wynn-Williams, C. G., Becklin, E. E. and Neugebauer, G. 'Infrared Studies of H II Regions and OH Sources', *Astrophys. J.* 187, 473–85 (1974).

30 Frogel, J. A. and Persson, S. E. 'Infrared Sources in Sharpless 228', *Astrophys. J.* 186, 207–10 (1973).

31 Kleinmann, D. E. and Wright, E. L. 'A New Infrared Source in M 17', *Astrophys. J.* 185, L131–L133 (1973).

32 Aitken, D. K. and Jones, B. 'Observations of Ne II in the Compact H II Region G $333 \cdot 6$–$0 \cdot 2$', *Mon. Not. Roy. Astron. Soc.* 167, 11P–15P (1974).

33 Allen, D. A., Strom, K.M., Grasdalen, G. L., Strom, S. E. and

Merrill, K. M. 'Haro 13a: a Luminous, Heavily Obscured Star in Orion?', *Mon. Not. Roy. Astron. Soc.* In Press. (1975).

34 Encrenaz, P. J. 'A New Source of Intense Molecular Emission in the Rho Ophiuchi Complex', *Astrophys. J.* 189, L135–L136 (1974).

CHAPTER 10

1 Connes, J., Connes, P. and Maillard, J. P. *Atlas des Spectres dans le Proche Infrarouge de Vénus, Mars, Jupiter et Saturne,* Centre National de la Recherche Scientifique, Paris (1969).

2 Connes, P., Connes, J., Benedict, W. S. and Kaplan, L. D. 'Traces of HCl and HF in the Atmosphere of Venus', *Astrophys. J.* 147, 1230–7 (1967).

3 Connes, P., Connes, J., Kaplan, L. D. and Benedict, W. S. 'Carbon Monoxide in the Venus Atmosphere', *Astrophys. J.* 152, 731–43 (1968).

4 Beer, R. and Taylor, F. W. 'The Abundance of CH_3D and the D/H Ratio in Jupiter', *Astrophys. J.* 179, 309–27 (1973).

5 Beer, R., Norton, R. H. and Martonchik, J. V. 'Absorption by Venus in the 3–4 Micron Region', *Astrophys. J.* 168, L121–L124 (1971).

6 Connes, P., Connes, J., Bouique, R., Querci, M., Chauville, J. and Querci, F. 'Sur les Sceptres d'étoiles Rouge M et C entre 4,000 et, 9,000 cm^{-1} (2,5 à 1,1 μ)', *Annals d'Astrophysique*, 31, 485–92 (1968).
   Connes, P. and Michel, G. 'High-resolution Fourier Spectra of Stars and Planets', *Astrophys. J.* 190, L29–L32 (1974).

7 Johnson, H. L. 'The Infrared Spectrum of the NML Cygnus Object,' *Astrophys. J.* 154, L125–L129 (1968).

8 Gillett, F. C., Stein, W. A. and Low, F. J. 'The Spectrum of NML Cygnus from 2·8 to 5·6 Microns', *Astrophys. J.* 153, L185–L188 (1968).

9 Thompson, R. I., Schnopper, H. W., Mitchell, R. I. and Johnson, H. L. '1–4-micron Spectra of Four Carbon Stars and Sirius', *Astrophys. J.* 158, L55–L60 (1969).
   Thompson, R. I., Schnopper, H. W., Mitchell, R. I. and Johnson, H. L. '1–4-micron Spectra of Four M Stars and Alpha Tauri', ibid, L117–L122 (1969).
   Forbes, F. F., Stonaker, W. F. and Johnson, H. L. 'Stellar and Planetary Spectra in the Infrared from 1·35 to 4·10 Microns', *Astron. J.* 75, 158–164 (1970).
   Johnson, H. L. and Méndez, M. L. and 'Infrared Spectra for 32 Stars' *Astron. J.* 75, 785–817 (1970).
   Thompson, R. I. and Schnopper, H. W. 'Identification of Infrared CN Bands in the Spectra of Several Carbon Stars', *Astrophys. J.* 160, L97–L100 (1970).

10 Frogel, J. A. 'Water Absorption in the Infrared Spectrum of Long-period Variable Stars and Associated Microwave Emission', *Astrophys. J.* 162, L5–L9 (1970).

11 Beer, R., Hutchison, R. B., Norton, R. H. and Lambert, D. L. 'Astronomical Infrared Spectroscopy with a Connes-type Interfero-

meter, III. Alpha Orionis, 2,600–3,450 cm^{-1}', *Astrophys. J.* 172, 89–115 (1972).

12 Geballe, T. R., Wollman, E. R. and Rank, D. M. 'Observations of Carbon Monoxide at 4·7 Microns in IRC+10216, VY Canis Majoris and NML Cygni', *Astrophys. J.* 183, 499–504 (1973).

13 Rank, D. M., Geballe, T. R. and Wollman, E. R. 'Detection of ^{17}O in IRC+10216', *Astrophys. J.* 187, L111–L112 (1974).

14 Johnson, H. L., Thompson, R. I., Forbes, F. F. and Steinmetz, D. L. 'The Infrared Spectrum of Alpha Herculis from 4000 to 4800 cm^{-1}', *Publ. Astron. Soc. Pacific* 84, 775–8 (1972).

Thompson, R. I., Johnson, H. L., Forbes, F. F. and Steinmetz, D. L. 'The Infrared Spectrum of Alpha Herculis from 5700 to 6700 cm^{-1}', ibid, 779–83 (1972).

Johnson, H. L., Thompson, R. I., Forbes, F. F. and Steinmetz, D. L. 'The Infrared Spectrum of χ Cygni from 4,000 to 6,700 cm^{-1}', ibid 85, 179–86 (1973).

Thompson, R. I., Johnson, H. L., Forbes, F. F. and Steinmetz, D. L. 'The Infrared Spectra of Two Carbon Stars from 4,000 to 6,700 cm^{-1}', ibid, 643–52 (1973).

15 Gillett, F. C. and Stein, W. A. 'Detection of the 12·8-micron Ne$^+$ Emission Line from the Planetary Nebula IC 418', *Astrophys. J.* 155, L97–L100 (1969).

16 Aitken, D. K. and Jones, B. 'Observations of Ne II in the Compact H II Region G 333·6–0·2', *Mon. Not. Roy. Astron. Soc.* 167, 11P–15P (1974).

17 Rank, D. M., Holtz, J. Z., Geballe, T. R. and Townes, C. H. 'Detection of 10·5-micron Line Emission from the Planetary Nebula NGC 7027', *Astrophys. J.* 161, L185–L189 (1970).

Holtz, J. Z., Geballe, T. R. and Rank, D. M. 'Infrared Line Emission from Planetary Nebulae', *Astrophys. J.* 164, L29–L33 (1971).

Geballe, T. R. and Rank, D. M. 'Observation of 9·0-micron Line Emission from Ar III in NGC 7027 and NGC 6572', *Astrophys. J.* 182, L113–L116 (1973).

18 Gillett, F. C., Forrest, W. J. and Merrill, K. M. '8–13-micron Spectra of NGC 7027, BD +30° 3639, and NGC 6572', *Astrophys. J.* 183, 87–93 (1973).

19 Aitken, D. K. and Jones, B. 'Some Features of the Infra-red Spectrum of NGC 7027 and an Estimate of its Sulphur Abundance', *Mon. Not. Roy. Astron. Soc.* 165, 363–8 (1973).

20 Forbes, F. F. 'The Infrared Polarization of the Infrared Star in Cygnus',
21 *Astrophys. J.* 147, 1226–9 (1967).

Hashimoto, J-i., Maihora, T., Okuda, H. and Sato, S. 'Infrared Polarization of the Peculiar M-type Variable VY Canis Majoris', *Publ. Astron. Soc. Japan* 22, 335–40 (1970).

22 Shawl, S. J. and Zellner, B. 'Polarization of IRC +10216', *Astrophys. J.* 162, L19–L20 (1970).

23 Dombrovskii, V. A. and Khozov, G. V. 'Photometric and Polarimetric Study of Infrared Stars in the Visible and Infrared Spectral Regions', *Astrophys.* 8, 1–7 (1974. Russian original dated 1972).

24 Dyck, H. M., Forbes, F. F. and Shawl, S. J. 'Polarimetry of Red and Infrared Stars at 1 to 4 Microns', *Astron J.* 76, 901–15 (1971).

25 Capps, R. W. and Dyck, H. M. 'The Measurement of Polarized 10-micron Radiation from Cool Stars with Circumstellar Shells', *Astrophys. J.* 175, 693–7 (1972).

26 Breger, M. and Hardorp, J. 'Infrared Polarimetry of Very Young Objects including the Becklin–Neugebauer Source', *Astrophys. J.* 183, L77–L79 (1973).

27 Loer, S. J., Allen, D. A. and Dyck, H. M. '2·2- and 3·5-micron Polarization Measurement of the Becklin–Neugebauer Object in the Orion Nebula', *Astrophys. J.* 183, L97–L98 (1973).

28 Dyck, H. M., Capps, R. W., Forrest, W. J. and Gillett, F. C. 'Discovery of Large 10-micron Linear Polarization of the Becklin–Neugebauer Source in the Orion Nebula', *Astrophys. J.* 183, L99–L102 (1973).

29 Dyck, H. M., Capps, R. W. and Beichmann, C. A. 'Infrared Polarization of the Galactic Nucleus', *Astrophys. J.* 188, L103–L104 (1974).

30 Dyck, H. M., Forrest, W. J., Gillett, F. C., Stein, W. A., Gehrz, R. D., Woolf, N. J. and Shawl, S. J. 'Visual Intrinsic Polarization and Infrared Excess of Cool Stars', *Astrophys. J.* 165, 57–66 (1971).

31 Gehrels, T. 'Wavelength Dependence of Polarization. XXVII. Interstellar Polarization from 0·22 to 2·2 μm', *Astron. J.* 79, 590–3 (1974).

INDEX

A

Accuracy 39
Aircraft observations 50
Albedo 63-66, 68, 69
Ammonia 65
Amplifiers 34, 35
Analyser (polarizing) 191
Aperture 32
Asteroids 67-70
Atmospheric extinction 37
Atmospheric transmission 16, 23, 24
Atmospheric windows 24, 25

B

Background 155, 157
Balloon observations 50
Bamberga 69
Beamsplitter 30
Becklin's object (Becklin's star; Becklin-Neugebauer source; **BN**) 166-178, 192, 193
Bennett, comet, 73, 81, 82
Be stars 41, 98-106
Black body 13, 17, 58
Black-body curve – *see* Planck function
Bolometer 21-23
Boulders (in hot spots) 58
Bremsstrahlung – *see* Free-free radiation
Brightness temperature 57
B stars 41, 97
Bug Nebula = NGC 6302

C

Calibration 38, 39
Calorific rays 11
Carbonate dust 137, 190
Carbon monoxide 83, 84
CD —56° 8032 113

Centaurus A = NGC 5128
Ceres 66, 69
Chopping 27-31
Circumstellar dust 78, 81-86, 100-107, 112 113, 121, 123, 126, 164, 189, 190, 193, 196
CIT sources 43, 79, 85, 158, 179, 193
Cocoon stars 78, 79
Colour temperature 57
Comet 1965f 70, 73
Comet 1969i 73, 80, 81
Comet 1973f 74
Cometary nebulae 125
Compact H II regions 132, 135-139
Compact galaxies 144, 145
Cone Nebula 174, 175
Crab Nebula 114
Crab pulsar 114
Cygnus A 144
Cygnus X-3 115
Cygnus OB2 no 12 163, 172
3C 120 144
3C 273 142, 147

D

DAA 6 173-176, 193
Dark nebulae 159-163
Deuterium 187
Dewar 31-33
Doping 23
Dust 71, 73, 79, 81-86, 100-107, 112, 113, 116, 117, 119, 121, 123, 126, 131-135, 137, 150, 152, 155, 160, 161, 163, 164, 167, 168, 170, 189, 190, 192, 193, 196
Dwarf stars 92

E

Eclipse, lunar 19, 55, 56

Effective wavelength 25
Electromagnetic spectrum 11, 12
Emission of sky 25
Emission-line galaxies 147
Emission-line stars 41, 94, 95, 97-113, 117-128
Emissivity 17, 18, 58, 81, 200
Ethane 65
Extinction, atmospheric 37, 38
Extinction, interstellar – see Reddening

F

Far infrared 49, 50
Far-infrared background 155, 157
Felgett, P. B. 40
Field lens 32
Filter 19
Filter-wheel spectrometer 184-186
Fish's Mouth Nebula 167
Flux 13-15, 38, 39
Footprint Nebula = Minkowski 1-92
Forbidden-line spectra 100, 102, 108, 110
Fortuna 69
Free-bound radiation 90
Free-free radiation 88-90, 98-100, 105, 112, 128, 132, 149, 167, 177, 196
Frequency 12, 13

G

G 333·6 —0·2 139, 176, 190
Galactic centre 152-156, 172, 179, 193
Galaxies 140-157
Galilean satellites 63-65
Germanium bolometer 21-23
Giant stars 83
Globular clusters 95
Golay cell 20
Graphite 84, 86, 92, 113, 194
Grating spectrometer 184-186

H

H II regions 108, 117, 131, 139, 154, 155, 167, 176
Haro 2-3 137
Haro 2-6 137
Haro 13a = IC 430
Haro-Chavira objects 93
HD 45677 101, 102, 106

HD 51585 101
HD 77581 115
HD 101584 83
HD 153919 114
Herbig-Haro objects 128-130
Herschel, Sir W. 9-11
Herschel 36 – see Messier 8
H_2O emission 86
Horsehead Nebula 167
Hot spots (lunar) 55-57
Hourglass – see Messier 8

I

Iapetus 66
IC 418 189, 190
IC 430 176-178
IC 1795 135, 172
IC 2087 163
Ice 64, 65, 168-173, 178, 194
Ic variables – see SRc variables
Ikeya-Seki, comet 70-72
Indium antimonide detectors 195
Inertia – see Thermal inertia
Infrared excess (definition) 76
Infrared catalogue – see IRC
Infrared telescope (62-inch) 41, 42, 96
Integration 35, 37
Interstellar dust 40, 134, 139, 153, 162-164, 190
IRC 42, 43, 76, 97, 98, 126, 131, 137, 191
IRC —20 385 95
IRC +10 216 85, 90, 91, 179, 188, 192, 193
IRC +10 420 105
IRC +20 052 193
IRC +30 021 193
IRC +40 013 = Messier 31
IRC +40 091 = Lk Hα101
IRC +40 430 = Cygnus OB2 no 12
IRC +50 137 86
Isotopes 188

J

Johnson, H. L. 24, 40, 41
Jupiter 60-63, 66, 187

K

Kleinmann-Low nebula (**KL**) 134, 166-170

Kohoutek 3-50 137
Kohoutek, comet 74

L

Lambert's law 54, 58, 60, 70
Lead sulphide (PbS) detectors 21, 22, 33, 40, 194, 195
Leighton, R. B. 41-43
L Hα25 120
Limb darkening 51, 61, 62, 96
Liquid gases 22, 33
Lk Hα101 125, 126
Lk Hα208 122, 123
Lk Hα269 = V1057 Cygni
Low, F. J. 23, 44, 45
Low bolometer – see Bolometer

M

Maffei objects 151
Magellanic Clouds 155
Magnitudes 37, 38
Markarian galaxies 146
Markarian 9 146
Markarian 231 142, 146
Mars 39, 60, 66, 187
Mass loss 84, 105, 110
Mauna Kea Observatory 47, 48
Menzel 3 109, 110
Mercury 58, 59, 66
Messier 1 – see Crab Nebula
Messier 8 135
Messier 15 95
Messier 17 131, 132, 176, 179
Messier 20 131
Messier 22 95
Messier 31 140
Messier 42, 43 – see Orion Nebula
Messier 51 144, 145
Messier 77 141, 148-150
Messier 82 141, 143, 144, 148, 150
Messier 87 144
Methane 61, 65, 66
Meteors 74, 75
Michelson interferometer 184-186
Micrometric diameters of asteroids 67, 68
Minkowski 1-78 137
Minkowski 1-82 160, 161
Minkowski 1-92 130, 131

Minkowski 2-9 105
Minkowski 4-18 112
Molecular absorption 52, 61, 83, 86, 94, 186-188
Molecular clouds 154, 167, 168, 175, 176
Moon 19, 52-59, 66
Mount Lemmon Observatory 45-47, 49, 184
MWC 17 101
MWC 297 126
MWC 645 103
MWC 819 103
MWC 922 101, 103
MWC 939 103
MWC 1080 120

N

Nebulium 108
Nebulous stars 117
Neon (12·8 m emission line) 189
Neptune 66, 67
Neugebauer, G. 41-43
Ney, E. P. 45
Ney-Allen source (NA) 134, 166
NGC 253 143, 150
NGC 1068 141, 148-150
NGC 1554-5 121
NGC 1579 126
NGC 2024 132
NGC 2237-9 176
NGC 2261 118, 119, 123
NGC 2264 117, 121, 122, 173-175
NGC 4151 141, 148, 149
NGC 5128 144, 146
NGC 5194-5 144
NGC 6302 137, 138
NGC 6357 133, 179
NGC 6572 190
NGC 7009 190
NGC 7027 135-137, 189, 190
NGC 7538 176
NGC 7552 143
NGC 7582 143
NML Cygni 43, 76-79, 83, 86, 179, 188, 191-193
NML Cyg type stars 86-88
NML Tauri 43, 76-79, 191, 192
NML Tau type stars 86-90, 105
Nodding 27, 29

Nova Serpentis (1970) 113

O

Occultation, by moon 91
Occultation, involving asteroids 68
Offset guide 30, 31
OH sources 84, 86, 167, 176
Ophiuchus dark cloud 163, 178
Optical pyrometer 13
Optical thickness 73
Orion Nebula 122, 132-135, 166-170, 176, 192

P

Phobos 60
Phosphine 65
Photographic infrared 11, 13, 21
Photography in infrared 194
Photometer 30, 31
Photometry 21, 24, 25, 183, 184, 195
Planck function 13-17, 198-200
Planetary nebulae 98, 100, 106-110, 113
Pluto 66, 67
Polarization (infrared) 170, 183, 190, 194
Preamplifier 35
Protostars 166, 170-173
Proxima Centauri 92
Pulsar – see Crab pulsar
Pyrometer 13

Q

Quasars 147

R

Radiant heat 13
Radio galaxies 144, 146, 147, 149
Rayleigh-Jeans approximation 17, 89, 199
Rectification 35
Reddening 40, 41, 95, 97, 106, 115, 121, 151, 153, 161-164, 167-178, 196
Reference signal 35
Rings of Saturn 65
Rocket observations 50, 74
Rosette source 176
Rosse, 4th Earl 19

S

Sagittarius A 153
Sagittarius B 154
Satellites, of Mars 60
    of Jupiter 63-65
    of Saturn 66
Saturn 65, 66, 187
Scattering 25
Scorpio X-1 114, 115
Seeing 26
Seyfert galaxies 148
Sharpless 228 176
Sharpless 235 = Minkowski 1-82
Sharpless 269 176
Silicate dust 73, 74, 80-86, 88, 94, 103, 104, 106, 107, 134, 135, 168-173, 175, 176, 178, 188, 189, 192-194
Silicon carbide 86, 188
Sky noise 26
Smyth, C. P. 19
Spectrum, electromagnetic 11, 12
Spectrum, infrared 60, 61, 64, 65, 83, 137, 168-171, 183-190, 195, 196
SRc variables 82, 83
Standard stars 37
Steam absorption in stars 187, 188
Stefan's law 16, 199, 200
Stellar planetary nebulae 108
Subdwarf stars 91
Sun 11, 24, 51, 52, 66
Supergiant stars 83, 91, 110
Supernova (in NGC 5253) 114
Survey telescope (62-inch) 41, 42, 96
Symbiotic stars 93, 94, 102

T

Taurus dark clouds 163
Tenerife 19, 48, 49
Thermal anomalies – see Hot spots
Thermal inertia 54
Thermal inertia of various bodies 54, 55, 59, 60, 64, 70
Thermocouple 18
Thermopile 18, 19
Titan 66
Transmission, atmospheric 16
Trapezium cluster = θ^1 Orionis

U

United States Air Force (USAF)

survey 159, 175, 176
Uranus 66, 67

V

Variability (infrared) 95, 96, 148-151
Velghe 2-45 112
Venus 59, 60, 66, 187
Vesta 67, 69
Vidicons 194

W

Walker 90 = L H 25
Water cell 19
Westerhout 3 135, 172, 176
Westerhout 49, 51 and 75 176
Whitford, A. E. 40
Wien's law 16, 57, 71, 200
Winchester 69
Window 31
Wolf-Rayet stars 95, 98, 111, 113
Woolf, N. J. 45

X

X-ray stars 114, 115

Z

Zodiacal light 75
Zwicky galaxies 142, 144, 145

Stars, arranged by constellation
Z And 94
RW And 79
R Aqr 94, 95
ϵ Aur 44, 93
ψ^1 Aur 83
α Boo 81, 179
Z CMa 120, 124, 126
VY CMa 79-81, 179, 192-194
η Car 102-105, 179
KN Cas 83, 94
   Cen C (Proxima) 92
W Cep 94
VV Cep stars 94, 96
o Cet 81, 83, 179
R CrA group 129, 130
R CrB 91, 92
γ Cru 179
DG Cyg 79

V 1016 Cyg 95
V 1057 Cyg 126, 127
V 1329 Cyg 95
30 Dor 131
R Dor 80, 81, 179
WY Gem 83, 94
β Gru 179
α Her 179
89 Her 83
RU Her 79
AC Her 91
MW Her 79
W Hya 80, 81, 179
BL Lac 142, 147
BL Lac objects 147
α Leo 44
R Leo 179
17 Lep 83, 84
R Lep 86
α Lyr 77
R Mon 117-119
S Mon 117, 173
θ Mus 112
ζ Oph 193
α Ori 42, 43, 76, 81, 93, 179, 188
γ Ori 44
NF ζ Ori 132
θ^1 Ori 134, 135, 166, 168
FU Ori 126, 127
HK Ori 120
NU Ori 193
S Per 80, 81, 83
19 (TX) Psc 85
L_2 Pup 193
RX Pup 94
FG Sge 110
ν Sgr 92
VX Sgr 83, 193
RY Sgr 92
V348 Sgr 113
α Sco 94, 179
WX Ser 79
α Tau 77, 81
T Tau 117, 119
T Tau stars 108, 117, 123, 127, 128, 167
RV Tau stars 91
RY Tau 119
IK Tau = NML Tauri
γ^2 Vel 112